T0396585

Atlas of the Huai River Basin Water Environment: Digestive Cancer Mortality

Gonghuan Yang • Dafang Zhuang
Editors

Atlas of the Huai River Basin Water Environment: Digestive Cancer Mortality

Editors
Gonghuan Yang
Institute of Basic Medical Sciences
Chinese Academy of Medical Sciences
Beijing
People's Republic of China

Dafang Zhuang
Institute of Geographic Sciences
and Natural Resources Research
Chinese Academy of Sciences
Beijing
People's Republic of China

Sponsored by the Eleventh Five-Year Science and Technology Support Program (The Correlation between Cancer and the Huai River Water Environment, 2006BAI19B03)

ISBN 978-94-017-8618-8 ISBN 978-94-017-8619-5 (eBook)
DOI 10.1007/978-94-017-8619-5
Springer Dordrecht Heidelberg New York London

Library of Congress Control Number: 2014935245

Printed on acid-free paper

Springer is part of Springer Science+Business Media (www.springer.com)

Preface

The *Atlas of the Huai River Basin Water Environment: Digestive Cancer Mortality* is one outcome of the "*Evaluation of the Correlation between Cancer and the Huai River Water Environment*", a study conducted under the Eleventh Five-Year Science and Technology Support Program. The atlas is an important product of the project which describes changes in the water environment and the causes of death of the local population in the Huai River Basin over the past 30 years. This is done through a spatial reanalysis of existing monitoring data, with a particular focus on deaths from digestive cancers.

Researchers from the Chinese Center for Disease Control and Prevention and the Institute of Geographic Sciences and Natural Resources Research of the Chinese Academy of Science completed this project using routine monitoring data for water quality and cause of death surveillance data. These two sets of data were collected independently by the Ministry of Environmental Protection and the Ministry of Health, respectively.

When water monitoring data for each area of the river basin are mapped by year for the main stream of the river and for primary and secondary tributaries and lakes, the distribution and changes in water pollution are clear. Furthermore, a review of variation in trends in the causes of death in the Huai River Basin over the past 30 years shows that the areas which were the most seriously polluted for the longest time were precisely the areas with the highest increase in digestive cancer deaths. The increase was several times than that of the national average increase for the respective cancers. Spatial analysis shows a high level of correspondence between the seriously polluted areas and areas with high mortality from cancer. This is the most important finding of the atlas.

Although these results do not explain how water pollution causes cancer, which is a question to be answered by pathogenesis studies, they convincingly demonstrate that there is a certain correlation between the two. Water pollution not only affects the environment that people live in, but also has a serious adverse impact on their health, particularly through its role in promoting the occurrence and development of digestive cancers.

In recent years, there have been many media reports about how environmental pollution causes cancer and other diseases, and these have attracted considerable attention. "Comprehensive monitoring of environment and health" has also become a topic of common concern for the environment and health ministries. But although regular environmental and health monitoring provide information on the environment and on health, respectively, it is still not at all clear how a feasible, integrated environmental and health monitoring system can be created. This project establishes a database for environment and health in the Huai River Basin based on relevant survey and monitoring data. Routine monitoring data are used for spatial analysis and for identifying the correlation between environment and health problems. Our work constitutes an important attempt at evaluating environmental quality and providing an early warning of health problems that may arise. This study will contribute to further research on the impact of environmental pollutants on health and also to the implementation of comprehensive environment and health monitoring.

The best policies are those that are based on scientific evidence. This atlas convincingly proves the correlation between environmental pollution and health using objective data. We hope that the Atlas can help environmental experts, health experts and government at all levels to understand the history and current status of water pollution in Huai River Basin, and the implications of this for the current and future health status of the local population. In addition, the Atlas provides concrete evidence for experts conducting in-depth research on environment and health, for relevant levels of government involved in pollution management, and for other authorities concerned about pollution and health. We hope that this can enable more effective management of pollution, that the Huai River area can become safer, and that the people dwelling on its banks can live a healthier, happier life.

Beijing, People's Republic of China Gonghuan Yang

Contents

1 **Research on the Correlation between Cancer and the Huai River Water Environment** 1
Introduction 1
Natural Geography of the Huai River Basin 1
Socio-economic Status of the Huai River Basin 2
Environmental Pollution in the Huai River Basin 3
Methodology 3
Analysis of Surface Water Quality in the Huai River Basin 3
Distribution and Comparative Analysis of Mortality from Cancers in the Huai River Basin 5
Main Findings 6
Variation in the Water Environment of the Huai River Basin 6
Variation in Mortality from Digestive Cancers in the Huai River Basin 13
Conclusion 15
Appendix 17
Tables 22
References 26

2 **Parent Maps** 27

3 **Variation in the Main Stream Water Environment** 35

4 **Variation in the Water Environment of the Tributaries** 55

5 **Variation in the Water Environment of Lakes** 109

6 **Spatiotemporal Variation in the Frequency of Water Pollution** 111

7 **Age-Standardized Mortality Rate of Digestive Cancer** 123

Glossary 143

Index 145

Contributors

Editor in Chief

Gonghuan Yang, Dafang Zhuang

Associate Managing Editor

Hongyan Ren, Maigeng Zhou

Editorials

Cai Liqun; Fu Jingying; Fu Xinyu; Guo Yan; Hu Yisong; Huang Yaohuan; Ji Wei; Jiang Dong; Jiang Xiaoshan; Qi Xiaopeng; Ren Hongyan; Shi Xiaoming; Song Daping; Wan Xia; Wang Lijun; Xu Jianwei; Yang Gonghuan; Zhou Maigen; Zhuang Dafang

Institutions

Chinese Center for Disease Control and Prevention

Institute of Geographic Sciences and Natural Resources Research of the Chinese Academy of Science

Institute of Basic Medical Sciences of Chinese Academy of Medical Sciences/School of Basic Medicine of Peking Union Medical College

List of Figures

Fig. 1.A1 Water quality at Gan'an Bridge in Xinyang City changed between Grade II and Grade III from 1982–2009. Except in 1992 and 2004 when water quality was Grade IV, it was Grade II and Grade III in 92 % of years monitored, which indicates that water quality in the waterhead area is generally good. 17

Fig. 1.A2 The monitoring section at Wangjiaba in Fuyang City was not established until 2003 so there are only 7 years of monitoring results. Water quality in this section was Grade V in 2003–2004 and Grade IV in 2005–2009 17

Fig. 1.A3 Water quality in this section was Grade V or worse from 1995–1997, although it was Grade III in 1998 and Grade IV 2003. It began to improve from 2006 on and it was Grade III in 2009. Overall, the amount of years with water quality of Grade V or worse accounted for 50 % of all years for which there is monitoring data. 18

Fig. 1.A4 Although water quality of Dabukou section in Xi County improved in 1986, it started deteriorating from 1991 and it was Grade V or worse from 1995 to 1997. There was a second improvement in water quality in this section during 1998–2009 when it remained at Grade II. 18

Fig. 1.A5 Water quality in the Xiashankou section in Huainan City varied from Grade II to Grade VI or worse from 1982 to 2009. The years with Grade V or worse were 1987, 1991, and 2004–2005, which accounted for about 17 % of all for which there is monitoring data 18

Fig. 1.A6 The worst water quality (Grade V) was recorded in 2002 and 2005, accounting for one-fifth of all years. About 70 % of all years for which there is monitoring data reported water quality of Grade IV 19

Fig. 1.A7 Grade V or worse water quality occurred only once in the Huaibin Section in Xinyang City in 1986. After this, the water quality of this section remained at Grade II-III-IV 19

Fig. 1.A8 For the Shitoubu Section in Huainan City, Grade IV water quality was reported for most years, accounting for about 69 % of all years. Grade V appeared only in 2004 and Grade III was recorded for the other years (1999, 2001, and 2009). 19

Fig. 1.A9 Of the 14 years with monitoring results for the section at the entrance of the Guo River into Huai River, years with Grade IV water quality accounted for about 64 %. Grade V was observed only in 1996 and 2003 20

Fig. 1.A10 The Bengbuzha Section had water quality of Grade II and Grade III (48 %), Grade IV (35 %) and Grade V or worse (17 %) in various years, which shows that this segment of the mainstream of Huai River was once seriously polluted but has now shown some improvement. 20

Fig. 1.A11 Water quality in the Xiaoliuxiang Section in Chuzhou City was Grade IV from 1995–1997 and 2003–2007, or about 67 % of the time. Grade V or worse water quality was recorded in 1998 and 2002, and the best quality was Grade III during 2008–2009 20

Fig. 1.A12 The Xintie Bridge Section had water quality of Grade IV 47 % of the time and of Grade V or worse 41 % of the time (1989, 1991, 1993, 1996, 2001, 2002, and 2004), which shows that this segment of the Huai River was seriously polluted 21

Fig. 1.A13 Water quality for the Huaihe Bridge Section in Xuyi County, varied between Grade III and Grade IV 21

Fig. 1.A14 Water quality of Grade IV or better was mostly reported for the Mohekou Section and Grade V or worse water quality mainly appeared in 1989, 1991, 1993, 1996, 2001, 2002 and 2004. 21

Fig. 2.1 Topography 28

Fig. 2.2 Water system and water quality monitoring sections 29

Fig. 2.3 Groundwater quality 30

Fig. 2.4 Administrative divisions 31
Fig. 2.5 Population density (2004) 32
Fig. 2.6 Per Capita GDP (2005) 33
Fig. 2.7 Sample counties in different regions 34
Fig. 3.1 Water quality grades (1986–1995) 36
Fig. 3.2 Comparison of water quality grades (1986–1995) 37
Fig. 3.3 BOD concentration grades (1986–1995) 38
Fig. 3.4 COD concentration grades (1986–1995) 39
Fig. 3.5 Ammonia nitrogen concentration grades (1986–1995) 40
Fig. 3.6 Water quality grades (1995–2005) 41
Fig. 3.7 Comparison of water quality grades (1995–2005) 42
Fig. 3.8 BOD concentration grades (1995–2005) 43
Fig. 3.9 COD concentration grades (1995–2005) 44
Fig. 3.10 Ammonia nitrogen concentration grades (1995–2005) 45
Fig. 3.11 Water quality grades (2005–2009) 46
Fig. 3.12 Comparison of water quality grades (2005–2009) 47
Fig. 3.13 BOD concentration grades (2005–2009) 48
Fig. 3.14 COD concentration grades (2005–2009) 49
Fig. 3.15 Ammonia nitrogen concentration grades (2005–2009) 50
Fig. 3.16 Proportion of water quality grades (1986–2009) 51
Fig. 3.17 Proportion of BOD concentration grades (1982–2009) 52
Fig. 3.18 Proportion of COD concentration grades (1982–2009) 53
Fig. 3.19 Proportion of ammonia nitrogen concentration grades (1982–2009) 54
Fig. 4.1 Primary tributaries water quality grades (1997–2005) 56
Fig. 4.2 Comparison of primary tributaries water quality grades (1997–2005) 57
Fig. 4.3 Primary tributaries BOD concentration grades (1997–2005) 58
Fig. 4.4 Primary tributaries COD concentration grades (1997–2005) 59
Fig. 4.5 Primary tributaries ammonia nitrogen grades (1997–2005) 60
Fig. 4.6 Primary tributaries water quality grades (2005–2009) 61
Fig. 4.7 Comparison of primary tributaries water quality grades (2005–2009) 62
Fig. 4.8 Primary tributaries BOD concentration grades (2005–2009) 63
Fig. 4.9 Primary tributaries COD concentration grades (2005–2009) 64
Fig. 4.10 Primary tributaries ammonia nitrogen grades (2005–2009) 65
Fig. 4.11 Proportion of primary tributaries water quality grades (1997–2009) 66
Fig. 4.12 Proportion of primary tributaries BOD concentration grades (1997–2009) 67
Fig. 4.13 Proportion of primary tributaries COD concentration grades (1997–2009) 68
Fig. 4.14 Proportion of primary tributaries ammonia nitrogen grades (1997–2009) 69
Fig. 4.15 Secondary tributaries water quality grades (1997–2005) 70
Fig. 4.16 Secondary tributaries BOD concentration grades (1997–2005) 71
Fig. 4.17 Secondary tributaries COD concentration grades (1997–2005) 72
Fig. 4.18 Secondary tributaries ammonia nitrogen grades (1997–2005) 73
Fig. 4.19 Secondary tributaries water quality grades (2005–2009) 74
Fig. 4.20 Secondary tributaries BOD concentration grades (2005–2009) 75
Fig. 4.21 Secondary tributaries COD concentration grades (2005–2009) 76
Fig. 4.22 Secondary tributaries ammonia nitrogen grades (2005–2009) 77
Fig. 4.23 Proportion of secondary tributaries water quality grades (1997–2009) 78
Fig. 4.24 Proportion of secondary tributaries BOD concentration grades (1997–2009) 79
Fig. 4.25 Proportion of secondary tributaries COD concentration grades (1997–2009) 80
Fig. 4.26 Proportion of secondary tributaries ammonia nitrogen grades (1997–2009) 81
Fig. 4.27 Beijing-Hangzhou Canal water quality grades (1997–2005) 82
Fig. 4.28 Beijing-Hangzhou Canal BOD concentration grades (1997–2005) 83
Fig. 4.29 Beijing-Hangzhou Canal COD concentration grades (1997–2005) 84
Fig. 4.30 Beijing-Hangzhou Canal ammonia nitrogen grades (1997–2005) 85
Fig. 4.31 Beijing-Hangzhou Canal water quality grades (2005–2009) 86
Fig. 4.32 Beijing-Hangzhou Canal BOD concentration grades (2005–2009) 87
Fig. 4.33 Beijing-Hangzhou Canal COD concentration grade (2005–2009) 88
Fig. 4.34 Beijing-Hangzhou Canal ammonia nitrogen grades (2005–2009) 89
Fig. 4.35 Proportion of Beijing-Hangzhou Canal water quality grades (1997–2009) 90
Fig. 4.36 Proportion of Beijing-Hangzhou Canal BOD concentration grades (1997–2009) 91
Fig. 4.37 Proportion of Beijing-Hangzhou Canal COD concentration grade (1997–2009) 92
Fig. 4.38 Proportion of Beijing-Hangzhou Canal ammonia nitrogen grades (1997–2009) 93
Fig. 4.39 YiShuSi water system water quality grade (1997–2005) 94
Fig. 4.40 Comparison of other tributaries water quality grades (1997–2005) 95
Fig. 4.41 YiShuSi water system BOD concentration grades (1997–2005) 96
Fig. 4.42 YiShuSi water system COD concentration grade (1997–2005) 97
Fig. 4.43 YiShuSi water system ammonia nitrogen grades (1997–2005) 98

Fig. 4.44 YiShuSi water system water quality grades (2005–2009) 99
Fig. 4.45 Comparison of other tributaries water quality grades (2005–2009) 100
Fig. 4.46 YiShuSi water system BOD concentration grades (2005–2009) 101
Fig. 4.47 YiShuSi water system COD concentrations grade (2005–2009) 102
Fig. 4.48 YiShuSi water system ammonia nitrogen grades (2005–2009) 103
Fig. 4.49 Proportion of YiShuSi water system water quality grades (1997–2009) 104
Fig. 4.50 Proportion of YiShuSi water system BOD concentration grades (1997–2009) 105
Fig. 4.51 Proportion of YiShuSi water system COD concentration grades (1997–2009) 106
Fig. 4.52 Proportion of YiShuSi water system ammonia nitrogen grades (1997–2009) 107
Fig. 5.1 Water quality grades (1983–2009) 110
Fig. 6.1 Water quality in different regions (1997–2009) 112
Fig. 6.2 Frequency of water pollution (1997–2009) 113
Fig. 6.3 Frequency of water pollution (2001–2009) 114
Fig. 6.4 Frequency of water pollution (2005–2009) 115
Fig. 6.5 Frequency of BOD pollution (1997–2009) 116
Fig. 6.6 Frequency of BOD pollution (2005–2009) 117
Fig. 6.7 Frequency of COD pollution (1997–2009) 118
Fig. 6.8 Frequency of COD pollution (2005–2009) 119
Fig. 6.9 Frequency of ammonia nitrogen pollution (1997–2009) 120
Fig. 6.10 Frequency of ammonia nitrogen pollution (2005–2009) 121
Fig. 7.1 Age-standardised male mortality rate for digestive cancer (1973–1975) 124
Fig. 7.2 Age-standardised female mortality rate for digestive cancer (1973–1975) 125
Fig. 7.3 Age-standardised mortality rate for digestive cancer (2004–2006) 126
Fig. 7.4 Change in rates of age-adjusted cancer mortality, 1973–2006 127
Fig. 7.5 Age-standardised male mortality rate for liver cancer (1973–1975) 128
Fig. 7.6 Age-standardised female mortality rate for liver cancer (1973–1975) 129
Fig. 7.7 Age-standardised male mortality rate for liver cancer (2004–2006) 130
Fig. 7.8 Age-standardised female mortality rate for liver cancer (2004–2006) 131
Fig. 7.9 Change in rates of age-adjusted liver cancer mortality, 1973–2006 132
Fig. 7.10 Age-standardised male mortality rate for stomach cancer (1973–1975) 133
Fig. 7.11 Age-standardised female mortality rate for stomach cancer (1973–1975) 134
Fig. 7.12 Age-standardised male mortality rate for stomach cancer (2004–2006) 135
Fig. 7.13 Age-standardised female mortality rate for stomach cancer (2004–2006) 136
Fig. 7.14 Change in rates of age-adjusted stomach cancer mortality, 1973–2006 137
Fig. 7.15 Age-standardised male mortality rate for esophageal cancer (1973–1975) 138
Fig. 7.16 Age-standardised female mortality rate for esophageal cancer (1973–1975) 139
Fig. 7.17 Age-standardised male mortality rate for esophageal cancer (2004–2006) 140
Fig. 7.18 Age-standardised female mortality rate for esophageal cancer (2004–2006) 141
Fig. 7.19 Change in rates of age-adjusted esophageal cancer mortality, 1973–2006 142

Research on the Correlation between Cancer and the Huai River Water Environment

1

Introduction

Since the 1990s, dramatic population growth, the rapid development of industrial and agricultural production and the increase of township enterprises in the Huai River Basin have led to an ever increasing amount of domestic sewage, industrial waste water, urban rubbish, waste from mines and factories, medical waste, and pesticides and fertilizers from farmland being discharged into the river, mostly through ditches and as a result of rainfall. The Shaying River receives a daily average of 1.662 million tons of waste water from the 30 cities in Henan between Zhengzhou and Xiangcheng. Five counties and cities in the Fuyang area of Anhui Province discharge 0.138 million tons of waste water daily into the tributaries of the Kui, Xinbian and Sui rivers, causing serious pollution. Water pollution in the Huai River Basin has already become a major concern of the entire society.

Retrospective cause of death survey data for China in the 1970s show that there were low death rates from cancer in the upper and middle reaches of the Huai River at that time. With the exception of esophageal cancer, the mortality rate for digestive system cancers was lower than the national average, and so was the lung cancer mortality rate (The Editorial Committee 1979). However, since 2004, several media have reported the existence of "cancer villages" in these areas of Huai River Basin, and this has attracted the attention of various parties to the issue of high rates of cancer in the area.[1]

The Ministry of Science and Technology of the People's Republic of China supported the *Evaluation of the Correlation between Water Pollution and Cancer in the Huai River Basin*,[2] as part of its scientific research program on *Techniques for the Evaluation and Control of Environmental Impacts on Health*. The project studied the profile and characteristics of water pollution in the Huai River Basin, the spatial and temporal distribution of mortality from cancer, and the correlation between water pollution and mortality from digestive cancers in the area. This Atlas is one product of the project, which provides an objective, macro-level description of changes in water quality and local population health in the area over the last 30 years by using water monitoring and cause of death data.

Natural Geography of the Huai River Basin

The Huai River Basin is located in eastern China between the Yellow River Basin and the Yangtze River Basin (longitude E 111° 55′ 122° 45′, latitude N 30° 55′ 36° 20′). The Huai River is the natural geographical boundary between China's northern and southern climatic zones. As a transitional zone between the two, it is a warm temperate area with a north Asian, humid to semi-humid monsoon climate and four distinct seasons. The Huai River originates from Mount Funiu and Mount Tongbai in the west, and flows eastward into the Yellow Sea. To the south it is bordered by the Dabie Mountains, the Jianghuai hills, the Tongyang Canal and the south dike of the Rutai Canal, which separate the Huai River Basin from the Yangtze River Basin. To the north, the Huai River Basin is separated from the Yellow River Basin by the south dike of Yellow River and by Mount Tai. The west, southwest and northeast parts of the river basin are mountainous and hilly, and account for about 1/3 of the total watershed area. The remaining area is a large plain, which accounts for about 2/3 of the watershed, including lakes and depressions (see Topography, Fig. 2.1).

The total length of the river is 1,000 kilometers (km) and the total drop is 200 m. The upper reaches of the river, which are above Honghekou, cover a distance of 360 km, with a drop of 178 m and a gradient of 1/2,000. The middle reaches run from Honghekou to Zhongdu with an outlet out into

[1] Information Press of the State Council of People's Republic of China State, collection of press conference/2004, China Intercontinental Press.

[2] Ministry of the Science and Technology, *the Water Environment and Digestive Cancer Mortality in the Huai River Basin*, Project Number 2006BA119B03.

G. Yang, D. Zhuang (eds.), *Atlas of the Huai River Basin Water Environment: Digestive Cancer Mortality*,
DOI 10.1007/978-94-017-8619-5_1, © Springer Science+Business Media Dordrecht 2014

Hongze Lake. They cover a distance of 490 km, with a drop of 16 m and a gradient of 1/33,000. The lower reaches run from Zhongdu to Sanjiangying, covering a distance of 150 km, with a drop of 6 m and a gradient of 1/25,000. In addition to the channel which enters the Yangtze River, there is also the North Jiangsu irrigation canal and the Huaishuxin River, which diverts flood water to the Xinyi River.

There are many tributaries in the upper and middle reaches of the Huai River, of which 16 main branches cover a basin area larger than 2,000 km^2. Nine tributaries enter the river from the south, including the Shiguan, Pi, Dongfei and Chi rivers among others. All these tributaries, which originate in the Dabie Mountains and Jianghuai hills, run over hilly areas and have swift currents. Seven tributaries enter the Huai along the north bank, including the Hongru, Shaying, Guo and Kuisui rivers. With the exception of the hilly areas in the upper reaches of the Hongru, Shaying and Kuisui, all these tributaries run across plains. The Shaying River has the largest basin area, which is nearly 40,000 km^2 in size (see Water system and water quality monitoring sections, Fig. 2.2). There are several coastal channels east of the lower reaches of the Huai River, such as the Sheyang, Huangsha, Xinyang and Doulong harbors, which discharge water from the Lixia River and coastal areas. The basin area of the lower reaches is 25,000 km^2. The Yishusi water system, which flows through Shandong and Jiangsu Provinces, is composed of the Yi, Shu and Si rivers, and originates in the Yimeng Mountains in Shandong Province, with a total basin area of more than 16,000 km^2.

The groundwater quality map for the Huai River Basin (see Ground-water quality, Fig. 2.3) is derived from the Groundwater Quality Evaluation and Ranking in the *Atlas of China's Groundwater Resources and Environment* (Zhang and Li 2004). The groundwater at the upper reaches of the Huai River Basin and on the south bank is of good quality, and can be used directly for drinking water. The groundwater on the north bank of the middle reaches and large parts of the lower reaches has to be properly treated before being used for drinking. The groundwater in the middle reaches of the tributaries on the north bank, the northwest part of Nansi Lake and the coastal areas of the lower reaches is not suitable for drinking, but can be used for industrial and agricultural purposes. The quality of the groundwater in the middle reaches of the Ying River, the south bank of the middle reaches of Huai River (Huainan and Bengbu), the Kui River and in Yancheng city is of poor quality and unfit to be used for any purpose.

Socio-economic Status of the Huai River Basin

The Huai River Basin occupies an extremely important position in China's economic and social development. The Huai River flows through five provinces: Henan, Hubei, Anhui, Shandong and Jiangsu, including 189 counties (county-level cities) in 40 prefectures. The basin area is 270,000 km^2 (see Administrative divisions, Fig. 2.4), with a total population of about 165 million (Song et al. 2011). The Huai River Basin has a high population density of 610 people/km^2. This is about 4.6 times (Song et al. 2011) the national average (134 people/km^2) for the same period, and the highest among the major river basins. The characteristics of the population distribution are shown in the population density map (see Population density (2004), Fig. 2.5).[3]

Per capita GDP in the Huai River Basin is lower than the national average, and it is a relatively poor area. There are also very obvious differences in per capita GDP between different counties in the area (see Per capita GDP (2005), Fig. 2.6). The Huai River Basin is one of China's important agricultural production areas, with 13.33 million hectares of arable land, accounting for 1/8 of China's total. There is a crisscrossing network of rivers and canals, with numerous reservoirs, ponds, lakes and depressions. These extensive bodies of water cover an area of more than 13.34 million hectares and are rich in aquatic life, including more than 100 kinds of fish. The Huai River Basin is one of China's important freshwater fishing areas.

There are about 50 kinds of mineral resources in the Huai River Basin. Coal, with a verified reserve of more than 70 billion tons, ranks first, and accounts for about 1/8 of China's total. The installed capacity of power plants is nearly 20 million kilowatts. The Huai River Basin is the most important energy supplier in east China. Coastal areas of northern Jiangsu have long been important salt producing areas, and large salt mines are found throughout Huainan and western Henan.

The Huai River Basin has a well developed transport system. Three railway lines, the Beijing-Shanghai railway, the Beijing-Kowloon railway and the Beijing-Guangzhou railway connect the north and south in the east, middle and west parts of the Basin. The Longhai railway winds across from east to west. Highway networks extend in all directions, with 14 national highways running through the Basin. The Beijing-Hangzhou Grand Canal and the main channel of the Huai River constitute the backbone of inland water transport. There is a large airport hub in Zhengzhou, and smaller ones in Kaifeng, Xuzhou, Lianyungang and Fuyang, as well as large sea ports in Lianyungang and Shijiugang.

[3] The map of population density was derived from spatial data supplied by the Data Center for Resources and Environment Sciences.

Environmental Pollution in the Huai River Basin

Since the 1970s, along with rapid social and economic development in the Huai River Basin, water pollution has become a new type of water-related disaster in the area. According to statistics, by the end of 1998, nearly 200 large water pollution incidents had occurred in the Huai River Basin; more than 10 of them in the main channel of Huai River alone. In February 1989, February 1992 and July 1994, large-scale water pollution incidents occurred in the main channel of the Huai, seriously affecting industrial and agricultural production and the daily life of local residents (Zhu 2004). For example, shortly after the "Huai River '97 Zero Hour Action" a discharge of accumulated wastewater in the Linyi area polluted the main stream of the river, resulting in the deterioration of drinking water quality in Xuzhou city and a shortage of drinking water for hundreds of thousands of residents (Zhu 2004).

The Huai River Basin has a large population, highly developed agriculture and intensive land cultivation. Soil erosion and nutrient loss are very serious and have caused sediment and contaminants to enter the water system, leading to a decrease in the capacity of the reservoirs and the deterioration of water quality year by year. This has seriously restricted social and economic development in the Huai River Basin (Gu et al. 2006).

Methodology

Analysis of Surface Water Quality in the Huai River Basin

Data Sources

The data on water quality from 1982 to 2009 used in this atlas are from the series of *China Environmental Quality Reports*,[4,5] published by the State Environmental Protection Administration from 1983 to 2008, and by the Ministry of Environmental Protection from 2009 to 2010. We have selected for further analysis data from state-controlled sections of the Huai River that are monitored for water quality,[6] including water quality grades and monitoring indicators. The number of sections for water system in Huai River Basin was larger during 1982–2002 than for 2003–2009 (see Water system and water quality monitoring sections, Fig. 2.2). The total remained at 86 from 2003, including 14 sections for the main stream (see Appendix) and 72 sections for tributaries. Five tables are attached that provide details regarding missing monitoring water quality data from 1982 to 2009.

[4]Editor in chief of the China's Environmental Monitoring Station, *National Environmental Quality Report*, the State Environmental Protection Administration (SEPA), 1982–2008.

[5]Compiled by the Ministry of Environmental Protection of People's Republic of China, National Environmental Quality Report, China Environmental Science Press, 2009–2010.

[6]Sections in Hubei Province were not included in this dataset.

Standards for Water Quality Classification

Despite missing observed values or concentration grades, water quality monitoring data were nonetheless listed in the *China Environmental Quality Reports*. In some other years, observed concentration values were directly recorded. We therefore coded unclassified water quality data for the other years according to national standards (GB3838-1988 and GB3838-2002). The main monitoring indicators for water quality classification were ammonia nitrogen/non-ionic ammonia, biochemical oxygen demand (BOD), chemical oxygen demand (COD) and volatile phenols. The specific concentration ranges (mg/L) are shown in Table 1.1.

Non-ionic ammonia refers to nitrogen (NH_3) existing in water as free ammonia; ammonia nitrogen refers to the sum of free ammonia and ammonium ions (NH_4 +) in water. Biochemical oxygen demand (BOD) is the amount of dissolved oxygen needed by aerobic biological organisms in a body of water to break down organic materials present in a given water sample. Chemical oxygen demand (COD) is the amount of oxygen consumed by contaminants (organic materials, nitrite, ferrite and sulfide) in a solution of boiling potassium dichromate. Volatile phenols usually include phenols with a boiling point below 230 °C, which are mostly highly toxic monophenols.

For the sake of easy presentation, water quality grades classified according to relevant standards are used to express the concentration range of the indicators. For example, Grade III represents a concentration range of 1.0–1.5 mg/L for ammonia nitrogen and so on.

Water Systems and Regions in the Huai River Basin

In order to conduct systematic spatial analysis, the branches of the Huai River were classified into the main stream, primary tributaries, secondary tributaries, the Yishusi water system, the Beijing-Hangzhou Grand Canal and lakes (see Water system and water quality monitoring sections, Fig. 2.2). At the same time, the river basin was divided into seven regions according to the type of terrain, the distribution of the water systems and the direction of the rivers. These seven regions are: the lower reaches in the east, the central eastern plain, the hilly and mountainous area in the west, the central western plain, the southern plain, the Nansi Lake basin and the Yishusi water system (see Sample counties in different regions, Fig. 2.7).

Table 1.1 Concentration criteria for main monitoring indicators of water quality authorized by National Standards for different periods

Monitoring indicators	National standard	Concentration criteria (mg/L)				
		Grade I	Grade II	Grade III	Grade IV	Grade V
Non-ionic ammonia ≤	GB3838-1988	0.02	0.02	0.02	0.2	0.2
Ammonia nitrogen ≤	GB3838-2002	0.15	0.5	1.0	1.5	2.0
Biochemical oxygen demand (BOD) ≤	GB3838-1988	3	3	4	6	10
	GB3838-2002	3	3	4	6	10
Chemical oxygen demand (COD) ≤	GB3838-1988	2	4	6	8	10
	GB3838-2002	2	4	6	10	15
Volatile phenols ≤	GB3838-1988	0.002	0.002	0.005	0.01	0.1
	GB3838-2002	0.002	0.002	0.005	0.01	0.1

Comparison of Water Quality Grades and Their Proportions

In order to visually display overall water quality for each monitoring section, we computed the ratio of the incidence of the observation of various water quality grades and the concentration of related indicators (BOD, COD, ammonia nitrogen) to the total number of observations for that section. Because surveillance data were missing for certain sections in certain years (see Tables 1.4, 1.5, and 1.6), when calculating the ratios for these sections we focused on years for which data were available and did not include years for which they were missing. At the same time, we calculated the ratios of the sections grouped by water quality grades (e.g. Grade II-III, Grade IV, or Grade V-VI) for the main stream, main tributaries, secondary tributaries, Beijing-Hangzhou Canal, Yishusi Water System, Lakes, and seven regions. We thus obtained a series of maps of these ratios for each monitoring section, for each water system, for the lakes, and for the seven regions.

In order to provide a more detailed description of the variation in water quality for each section, we divided data for the main stream into three time periods (1986–1995, 1995–2005, and 2005–2009), and data for the main tributaries, secondary tributaries, Beijing-Hangzhou Canal, and Yishusi Water System into two time periods (1997–2005 and 2005–2009) according to the particular characteristics of the water quality data collected during different time periods. We thus created a series of maps of water quality grades and related indicators of surveillance sections from 1982 to 2009.

In order to visually compare water quality across two selected years, we set up a comparative indicator for water quality, which is expressed as the difference between the water quality grades of the year concerned (e.g. 2009) and the reference year (e.g. 2005). If the water quality is unchanged, the difference in grade is zero; if water quality has deteriorated the difference is 1; if deterioration is significant, 2 or more. If water quality has improved, the difference is −1; and if significantly improved, −2 or less. On the basis of this, we created a series of maps comparing water quality grades over above periods.

Frequency of Water Pollution

For monitoring sections for which long-term water quality data were available, the ratio of the occurrence of water quality Grade V or worse (polluted water) to total observations (frequency) is used to reflect water quality. This ratio is defined as the frequency of water pollution (FWP) and calculated via the following formula.

$$FWP = Y_p / Y$$

Where Y_p is the occurrence of water quality Grade V or worse in the given monitoring period (Times); and Y is total observations (e.g. yearly or monthly observations). Here, grade V or worse can refer to water quality after comprehensive evaluation or to the concentration level of a single indicator. Because surveillance data were missing for certain sections for certain years, (see Tables 1.4, 1.5, 1.6, 1.7, and 1.8 for details), these years were not taken into account when we calculated these ratios.

In addition, the reverse process was employed to calculate the Frequency of Water Pollution during some periods because concentrations of the main indicators were missing in the surveillance data from all the monitoring stations from 1998 to 2000. Accordingly, we acquired FWP in the periods from 1997 to 2009, from 2001 to 2009, and from 2005 to 2009 in order to analyze spatial variation in the distribution of serious water pollution during these years.

Spatialization of FWP

FWP is the proportion of times that water quality of Grade V or worse occurred for a given section for a given period of time. If this indicator is used as an attribute of the monitoring section for the purposes of spatial interpolation, it can represent the frequency of the occurrence of water quality Grade V or worse within a specified geographical space. It can reflect the distribution of seriously polluted bodies of water within the drainage basin and the probability of serious pollution for areas without surface water. Moreover, changes in the frequency of pollution for different periods inform us about changes and trends in pollution over time.

FWP was spatialized using GIS spatial interpolation in order to show the spatial distribution and variation in patterns of water pollution in the Huai River Basin. Spatial interpolation was performed according to the following steps:

A. Linking monitoring data and the spatial attributes of surveillance sections

   Monitoring data were organized so that each record includes the name of the river, the name of the section, the water quality grade and the concentration for each of the indicators. The section name is used as the key variable to link the monitoring data with section points, so that spatial information is associated with information about other attributes.

B. Spatial interpolation Methodology

   Spatial interpolation was performed using ARCGIS geostatistical analyst. Commonly used interpolation methods include Kriging interpolation, inverse distance weighted (IDW) interpolation, polynomial interpolation and radial basis function interpolation. Kriging interpolation has the advantage of being unbiased, but it is greatly affected by exceptional data points; IDW interpolation has lower precision than Kriging interpolation, but is more robust. Due to insufficient water quality monitoring data, the spatial interpolation method selected should be both scientific and robust. Thus, we employed IDW interpolation (Tao 2010). Finally, frequencies were obtained for each water quality grade and for grades corresponding to concentrations of BOD, COD, and ammonia nitrogen/non-ionic ammonia at 1 km resolution.

Distribution and Comparative Analysis of Mortality from Cancers in the Huai River Basin

Population

Demographic data for all monitoring counties were the village household population data reported by the counties from 2004 to 2006, The county population is the sum of the individual village populations. Demographic data were cross-referenced against 2000 census data to check reliability of age and gender distributions.

Data Sources for Cancer Mortality

- 1973–1975 national cause of death survey[7]

  The first national cause of death survey was carried out from 1973–1975, and covered a population of 850 million in 29 provinces, autonomous regions and municipalities. All families were asked if a family member had died within the past 3 years, and if there had been a death, the following information was collected: age at death, and place and cause of death. The completion rate was 98.5 %. The distribution and level of mortality by cause of death, age, gender and place in 1973–1975 were listed. The authors selected data for the Huai River Basin region for further analysis.

- 2004–2005 cause of death data for the Huai River Basin

  Over the past 30 years, there was no record of the cause of death or other basic information regarding the health status of the population in the Huai River Basin. Therefore, it was difficult to understand the health status of the population, and in particular, to confirm the accuracy of the many media reports in 2004 regarding "cancer villages" in the area or assess the severity of the problem.

In 2005, the Chinese Center for Disease Control and Prevention (China CDC) carried out an epidemiological survey of cancer incidence and risk factors in three key areas of the Huai River Basin.[8] On this foundation, in 2007, the China CDC selected one county in each section of the main stream and each of the tributaries of the Huai River – a total of 14 counties – and conducted a 3-year retrospective cause of death survey for 2004–2006. This survey was also part of the National Cause of Death and Epidemiological Survey (MOH 2008). On the basis of this, the China CDC set up a comprehensive surveillance system for the cause of death, morbidity, births and birth defects. This surveillance system covered 12.64 million people or 8 % of the total population of the Huai River Basin. The 14 sample counties are distributed across the six water environment regions described above.

When local residents die in hospital, the physician fills out a medical death certificate. In cases where people die at home, a trained village health worker at township hospital uses a verbal autopsy questionnaire to interview family members and collect the relevant medical information from before the victim's death in order to assess the probable cause of death and fill out the certificate. The local CDC submits the medical death certificate using the national platform for reporting cause of death information. A report analyzing the causes of death of local residents in the Huai River Basin was completed annually by the China CDC (Chinese Center for Disease Control and Prevention 2007) indicate years.

Disease Classification

According to the International Standards for Disease Classification, causes of death include infectious diseases, neoplasms, endocrine disorders, neuropsychiatric diseases, circulatory system diseases, respiratory diseases, digestive system diseases, urinary and reproductive system disorders, obstetric diseases, injuries and unknown diseases. Neoplasms

[7]Office of Cancer Prevention and Treatment Research, the Ministry of Health, survey of China's deaths from cancers, the People's Health Publishing House, 1979 Beijing.

[8]Chinese Center for Disease Control and Prevention, Study report on key areas of cancer incidence in Huai River Basin and risk factors, 2006.03.30, internal data.

Table 1.2 ICD-10 for digestive cancer

Cause of death	ICD-10
Neoplasms	C00-D48
Digestive system neoplasms	C15-C20
Malignant neoplasm of the esophagus	C15
Malignant neoplasm of the stomach	C16
Malignant neoplasm of the liver and intrahepatic bile ducts	C22

are further divided into esophageal cancer, gastric cancer, liver cancer, colorectal cancer, lung cancer, breast cancer, cervical cancer, leukemia and other cancers. ICD-10 codes for digestive cancer, liver cancer, gastric cancer and esophageal cancer are shown in Table 1.2.

Analysis of Indicators

Annual average population: (population at the end of the previous year + population at the end of the current year × 2 + population at the end of the following year) / 4)

Cancer mortality rate: (deaths from cancers during the year / average population of the same year) × 100,000/100,000.

Standardized cancer mortality rate: using 2000 population census data to conduct standardization makes the mortality rates of different counties and different years comparable.

$$P^c = \sum\left(\frac{N_1}{N}\right)\rho 1$$

1973–1975 standardized cancer mortality rate

2004–2006 standardized cancer mortality rate

Variation in cancer mortality: (The standardized cancer mortality rate of the sample area in 2004–2006) minus (the cancer mortality rate of the sample area in 2004–2006) divided by the cancer mortality rate of the sample area in 1973–1975.

Main Findings

Variation in the Water Environment of the Huai River Basin

The Main Stream of the Huai River

- 1986–1995

  The water quality fluctuated significantly in most monitoring sections of the main stream during this period. It was observed that the water quality exceeded the standard (Grade V or worse) in some upstream monitoring sections such as Dabukou in Xi County and Huaibin Hydrologic Station and some midstream monitoring sections such as Xiashankou, Dajiangou, Bengbuzha, Xintie Bridge and Mohekou. The water quality clearly deteriorated in Wangjiaba–Xiashankou, Woheruhuaikou–Bengbuzha and Xintie Bridge – Mohekou reaches (See Comparison of water quality grade (1986, 1995), Fig. 3.2).

  In this period, Dajiangou (See BOD concentration grade (1986–1995), Fig. 3.3) was the only monitoring section where the biochemical oxygen demand (BOD) concentration exceeded the standard (>6 mg/L). Meanwhile, the chemical oxygen demand (COD) concentration exceeded the standard (>10 mg/L) in the Dabukou, Huaibin Hydrologic Station and Dajiangou monitoring sections (See COD concentration grade (1986–1995), Fig. 3.4). In addition, ammonia nitrogen exceeded the standard (>1.5 mg/L) mainly in the Dabukou, Xiashankou, Dajiangou, Bengbuzha, Xintie Bridge and Mohekou monitoring sections (See Ammonia nitrogen concentration grade (1986–1995), Fig. 3.5), among which Xiashankou, Dajiangou and Mohekou showed an excessively high concentration of ammonia nitrogen for 3 years. Ammonia nitrogen was the main pollution indicator for water quality in Anhui reach of the Huai River main stream during this period.
- 1995–2005

  The number of monitoring sections where the water quality exceeded the standard (Grade V or worse) increased significantly during this period; in addition to the Dabukou, Huaibin Hydrologic Station, Xiashankou, Dajiangou, Guoheruhuaikou, Bengbuzha, Xintie Bridge and Mohekou sections mentioned above, water quality exceeding the standard was also detected in Wangjiaba, Shitoubu, Xinchengkou, Xiaoliuxiang, etc. The water quality of the main stream showed three statuses (See Comparison of water quality grade (1995 and 2005), Fig. 3.7), "significantly improved" (Dabukou and Dajiangou monitoring sections), "improved" (Huaibin Hydrologic Station and Bengbuzha monitoring sections) and "unchanged" (Changtaiguan Gan'an Bridge, Shitoubu, Guoheruhuaikou, Mohekou, Xiaoliuxiang, Huaihe Bridge, etc.).

  During this period, the monitoring sections where the BOD concentration exceeded the standard (>6 mg/L) mainly included Dabukou, Wangjiaba, Xintie Bridge, Xiaoliuxiang (See BOD concentration grade (1995–2005), Fig. 3.8). However, the BOD concentration of these monitoring sections in other years and of other monitoring sections between 1995 and 2009 were all at Grade IV or below. The COD concentration exceeding the standard (>10 mg/L) was observed mainly in Dabukou, Huaibin Hydrologic Station, Woheruhuikou, Bengbuzha, Xintie Bridge, Mohekou (See COD concentration grade (1995–2005), Fig. 3.9). The monitoring sections where the ammonia nitrogen concentration exceeded the standard (>1.5 mg/L) mainly included Dabukou, Huaibin Hydrologic Station, Xiashankou, Shitoubu, Dajiangou, Xinchengkou, Xintie Bridge, Mohekou, Xiaoliuxiang and etc. (See Ammonia nitrogen concentration grade (1995–2005), Fig. 3.10)

- 2005–2009

  The water quality of the main stream improved significantly during this period (See Water quality grade (2005–2009), Fig. 3.11), and the water quality exceeded the standard (grade V or worse) only in a few monitoring sections such as Xiashankou, Dajiangou and Xinchengkou in 2005. Water quality improved or significantly improved in most monitoring sections of the main stream, including Dabukou, Huaibin Hydrologic Station, Xiashankou, Shitoubu, Dajiangou, Guoheruhuaikou, Bengbuzha, Xintie Bridge, Mohekou, Xiaoliu Bridge and Huaihe Bridge (See Comparison of water quality grade (2005, 2009), Fig. 3.12).

  During this period, except for the Wangjiaba and Xiaoliuxiang monitoring sections, where the BOD concentration exhibited a level of Grade IV, none of the other sections exceeded the standard (>6 mg/L, See BOD concentration grade (2005–2009), Fig. 3.13). Meanwhile, a COD concentration of Grade IV appeared only in the Wangjiaba monitoring section in 2005 (See COD concentration grade (2005–2009), Fig. 3.14); for the other sections, the COD concentration remained at a level of Grade II or Grade III on a long-term basis. In terms of the ammonia nitrogen concentration of each monitoring section in this period (See Ammonia nitrogen concentration grade (2005–2009), Fig. 3.15), except in Dajiangou and Xinchengkou where it exceeded the standard (>1.5 mg/L) in 2005, the other sections' remained at grade IV or below

- Water quality grades and indicator concentrations in the main stream during 1982–2009

  As indicated by the proportion of monitoring sections where the water quality exceeded the standard (Grade V or worse) in each year (See Proportion of water quality grade (1982–2009), Fig. 3.16), there was obvious pollution in the water of the main stream in 1990s, especially in 1991–1993 and 1996 when this proportion reached or exceeded 60 %. During 2007–2009 this proportion declined overall, although it fluctuated in 2001–2002 and 2004–2005 (in 2004 this proportion rebounded to 43 %). In addition, the proportion of the years in which the water quality exceeded the standard (Grade V or worse) was higher than 30 % in monitoring sections like Dabukou, Xiashankou, Dajiangou, Xintie Bridge and Mohekou, while in other monitoring sections the proportion was mostly less than 30 % (there was no observation in water quality exceeding the standard for the Changtaiguan Gan'an Bridge and Xuyi Huaihe Bridge sections in this period).

  In terms of the proportion of monitoring sections in which the BOD concentration exceeded the standard (>6 mg/L), this only occurred in 1994–1995, 2002 and 2004 and the monitoring sections involved were primarily Wangjiaba, Xiashankou, Dajiangou, Xintie Bridge, Mohekou and Xiaoliuxiang. (See Proportion of BOD concentration grade (1982–2009), Fig. 3.17)

  The COD concentration exceeded the standard (>10 mg/L) in some monitoring section in 1986 and 1993–1997, which mainly included Dabukou, Huaibin Hydrologic Station, Shitoubu, Guoheruhuaikou, Bengbuzha, Xintie Bridge, Mohekou (See Proportion of COD concentration grade (1982–2009), Fig. 3.18).

  The figures show that (Proportion of ammonia nitrogen concentration grade (1982–2009)) in 1987–1989 as well as 1991–2005 there were monitoring sections where the ammonia nitrogen concentration exceeded the standard (>1.5 mg/L), and especially during 1991–1996 the proportion of such monitoring sections was above 40 %. In the entire period 1982–2009, ammonia nitrogen concentration did not exceed the standard in only a few monitoring sections such as Changtaiguan Gan'an Bridge, Wangjiaba, Guoheruhuaikou, Xuyi Huaihe Bridge, etc. It could be seen that, between 1982 and 2009 the ammonia nitrogen concentration seriously exceeded the standard in the main stream, and this phenomenon occurred mainly in the monitoring sections in the Xiashankou-Xiaoliuxiang reach in the midstream of the Huai River.

Tributaries

This Atlas describes the features of water environment changes in the tributaries of the Huai River. Different parts are presented separately, including primary tributaries, secondary tributaries, the Beijing-Hangzhou Canal and the Yishu-Si River System.

Primary Tributary

- 1997–2005

  Except in a few monitoring sections such as Jiangji Hydrologic Station, Xiaowang Bridge and Qubeizha, water quality exceeding the standard appeared in all the other monitoring sections (See Primary tributaries water quality grade (1997–2005), Fig. 4.1). Among these sections, some showed a water quality exceeding the standard for 3 or more years, including Pipashan Bridge, Huaidianzha, Bantai, Shenqiuzhidian, Taolao, Jieshou, Luyi Fu Bridge, Bozhou, Huangkou, Lixin reach, downstream Mengcheng reach, Xinandukou, downstream Yingshang, Wulizha, Gonglu Bridge, Sixian Bali Bridge, Sixian Gonglu Bridge and Suzui; these above monitoring sections accounted for 72 % of all monitoring sections. The changes in water quality grade in primary tributaries can be classified into five types (See Comparison of primary tributaries water quality grade (1997, 2005), Fig. 4.2), namely, "unchanged" (Shenqiuzhidian in Yinghe River, Bozhou in Guohe River and Hengchuan Hydrologic Station), "significantly deteriorated" (Jieshou, Lixin in West Feihe River and Jiangji Hydrologic Station in Shihe River), "deteriorated" (downstream Yingshang and downstream Mengcheng), "significantly improved" (Pipashan Bridge in Xinyang) and "improved" (Luyi Fu Bridge in Guohe River). The reaches and monitoring sections classified as "deteriorated" or "significantly deteriorated" were mainly concentrated in the tributaries on the north bank of the Huai River in Anhui Province.

In terms of the BOD concentration grade for primary tributaries (1997–2005) the monitoring sections where the BOD concentration exceeded the standard (>6 mg/L) mainly included Pipashan Bridge, Huaidianzha, Bantai, Shenqiuzhidian, Taolao, Jieshou, Luyi Fu Bridge, Bozhou, downstream Mengcheng reach, Xinandukou, downstream Yingshang, Gonglu Bridge, Sixian Bali Bridge, Sixian Gonglu Bridge and Suzui, which accounted for 60 % of all monitoring sections in primary tributaries. Meanwhile, COD concentrations exceeding the standard (>10 mg/L) mainly appeared in monitoring sections including Pipashan Bridge, Bantai, Shenqiuzhidian, Taolao, Jieshou, Luyi Fu Bridge, Bozhou, Huangkou, downstream Mengcheng reach, Xinandukou, downstream Yingshang, Gonglu Bridge, Sixian Bali Bridge, Sixian Gonglu Bridge and Daqu, which accounted for 64 % of all monitoring sections in primary tributaries (See Primary tributaries COD concentration grade (1997–2005), Fig. 4.4). In addition, the monitoring sections where the ammonia nitrogen concentration exceeded the standard (>1.5 mg/L) mainly included Huaidianzha, Luyifuqiao, Shenqiuzhidian, Bozhou, Jieshou, Yongcheng Zhang Bridge, downstream Mengcheng reach, downstream Yingshang, East Feihe River Wulizha, Xinandukou, Bengbu Guzhen, Gonglu Bridge, Sixian Bali Bridge and Suzui, accounting for 56 % of all monitoring sections in primary tributaries (See Primary tributaries ammonia nitrogen concentration grade (1997–2005), Fig. 4.5).

- 2005–2009
  The number of monitoring sections with water quality exceeding the standard decreased in the period 1997–2005 (See Primary tributaries water quality grade (2005–2009), Fig. 4.6), and there was no water quality exceeding the standard in the Pipashan Bridge, Yongcheng Zhang Bridge, Xiaowang Bridge, Hengchuan Hydrologic Station, Lixin reach, Bengbu Guzhen and Qubeizha sections. Moreover, the number of monitoring sections which exceeded the standard of water quality for 3 or more years also decreased to 64 % of all monitoring sections in primary tributaries, such as Huaidianzha, Shenqiu Zhidian, Jieshou, Luyi Fu Bridge, Bozhou, downstream Mengcheng reach, downstream Yingshang, Gonglu Bridge, Sixian Bali Bridge, Daqu and Suzui. The changes in the water quality grades of primary tributaries can be classified into three types (See Comparison of primary tributaries water quality (2005, 2009), Fig. 4.7), namely, "significantly improved" (downstream Yingshang, Jiangji Hydrologic Station and Hengchuan Hydrologic Station), "improved" (Bantai, Taolao, Huangkou, Pipashan Bridge, Jieshou and downstream Mengcheng) and "unchanged" (worse than Grade V in Huaidian Sluice, Shenqiu Zhidian, downstream Yingshang and Bozhou sections; Grade V in Luyi Fu Bridge; Grade IV in Xiaowang Bridge, Yongcheng Zhang Bridge, Bengbu Guzhen and Lixin reach sections).

  In terms of changes in the BOD concentration, the monitoring sections which exceeded the standard (>6 mg/L) mainly included Huaidianzha, Luyi Fu Bridge, Shenqiu Zhidian, Jieshou, Taolao, Bozhou, downstream Mengcheng reach, East Feihe River Wulizha, Gonglu Bridge, Sixian Bali Bridge, Sixian Gonglu Bridge and Suzui, accounting for 48 % of all monitoring sections in primary tributaries (See Primary tributaries BOD concentration grade (2005–2009), Fig. 4.8). Meanwhile, COD concentrations exceeding the standard (>10 mg/L) were evident only in the Bozhou, Huangkou, downstream Mengcheng reach, Gonglu Bridge, Sixian Bali Bridge and Sixian Gonglu Bridge sections, accounting for 24 % of all monitoring sections in primary tributaries (See Primary tributaries COD concentration grade (2005–2009), Fig. 4.9). In addition, monitoring sections where water exceeded the standard for ammonia nitrogen (>1.5 mg/L) mainly included the Huaidianzha, Luyi Fu Bridge, Shenqiu Zhidian, Jieshou, Bozhou, downstream Yingshang, Xinandukou, downstream Mengcheng reach, Gonglu Bridge, Sixian Gonglu Bridge, Daqu and Suzui, accounting for 40 % of all monitoring sections in primary tributaries (See Primary tributaries ammonia nitrogen grade (2005–2009), Fig. 4.10).

  Compared with 1997–2005, there was a decrease in the number of monitoring sections in primary tributaries which exceeded water quality standards for BOD, COD or ammonia nitrogen. The concentrations of various pollution indicators which exceeded the standard also decreased, especially the COD, which decreased significantly. However, in certain monitoring sections, the BOD and ammonia nitrogen concentrations still remained at a high level which exceeded the standards.
- Water quality grades and indicator concentrations in 1997–2009
  The number of the monitoring sections which exceeded the standard for water quality (Grade V or worse) accounted for the highest proportion (higher than 70 %) in 2001, and, after that, the proportion gradually declined, reaching the lowest level (around 30 %) in 2009 (See Proportion of primary tributaries water

quality grade (1997–2009), Fig. 4.11). On the north bank of the main stream, the proportion of the years in which water quality exceeded the standard was lower than 50 % in some monitoring sections such as Yongcheng Zhang Bridge, Xiaowang Bridge, Lixin reach, Bengbu Guzhen and Daqu, and was higher than 50 % or even as high as 100 % (i.e., exceeding the standard every year) in the other monitoring sections. However, on the south bank of the main stream, except in Suzui and Gonglu Bridge, the proportion was lower than 50 % in almost all monitoring sections, among which Jiangji Hydrologic Station was the only monitoring section which never exceeded the standard.

The proportion of monitoring sections where the BOD concentration exceeded the standard (>6 mg/L) in each year fell between 25 and 45 % during this period (See Proportion of primary tributaries BOD concentration grade (1997–2009), Fig. 4.12), and showed an overall decline. However, the proportion of years in which the BOD concentration exceeded the standard was at relatively high levels in many monitoring sections including the Huaidianzha, Shenqiu Zhidian, Jieshou and downstream Yingshang monitoring sections in the Yinghe River, the Luyi Fu Bridge, Bozhou and Mengcheng monitoring sections in the Shahe River-Guohe River, the Sixian Bali Bridge monitoring section in the New Suihe River, the Sixian Gonglu Bridge in the New Bianhe River, the Suzui monitoring section in the Subei General Irrigation Canal and the Gongluqiao monitoring section in the Chihe River. Among these monitoring sections, the proportion of years in which the BOD concentration exceeded the standard was 100 % in Bozhou and Sixian Baliqiao, suggesting that the BOD concentrations of these monitoring sections exceeded the standard every year.

The proportion of monitoring sections where the COD concentration exceeded the standard (>10 mg/L) in each year decreased gradually (See Proportion of primary tributaries COD concentration grade (1997–2009), Fig. 4.13). The COD concentration met the standard in Huaidianzha, Lixin reach, Yongcheng Zhang Bridge, Xiaowang Brige, Qubeizha, East Feihe River Wulizha, Hengchuan Hydrologic Station, etc., and exceeded the standard to various extents in all the other monitoring sections.

The proportion of monitoring sections where the ammonia nitrogen concentration exceeded the standard (>1.5 mg/L) in each year fell between 20 and 50 % during this period (See Proportion of primary tributaries ammonia nitrogen grade (1997–2009), Fig. 4.14). The proportion of the years in which the ammonia nitrogen concentration exceeded the standard was at relatively high levels in monitoring sections that included Huaidianzha, Shenqiu Zhidian, Jieshou and downstream Yingshang monitoring sections in the Yinghe River, Luyi Fu Bridge, Bozhou and Mengcheng monitoring sections in the Shahe River-Guohe River, Sixian Bali Bridge and Daqu monitoring sections in the New Suihe River, and the Suzui monitoring section in the Subei General Irrigation Canal. Among these monitoring sections, the proportion was 100 % in Huaidianzha and Bozhou, which suggested that the ammonia nitrogen concentrations in these monitoring sections exceeded the standard every year.

Second Tributary

- 1997–2005

Except in Dachenzha and Zhumazha, water quality exceeded the standard (Grade V or worse) in almost all monitoring sections (See Secondary tributaries water quality grade (1997–2005), Fig. 4.15), and some monitoring sections showed a deteriorating water quality for 3 or more years, including Shakou, Lifenzha, Xihua Dawangzhuang, Dongsunying, Baogongmiao, Liuzhaicunhou, Niqiu, downstream Linquan reach, Funan, Xuzhuang, Maqiao, Huangqiao, Yangzhuang, Yigoudaqiao, Tianchang Chemical Plant and Xiyingqiao, accounting for 73 % of all monitoring sections in secondary tributaries.

During this period, monitoring sections where the BOD concentration exceeded the standard (>6 mg/L) mainly included Xihua Dawangzhuang, Dongsunying, Lifenzha, Baogongmiao, downstream Linquan reach, Liuzhaicunhou, Niqiu, Maoqiao, Funan, Xuzhuang, Huangqiao, Yangzhuang and Tianchang Chemical Plant, accounting for 59 % of all monitoring sections in secondary tributaries (See Secondary tributaries BOD concentration grade (1997–2005), Fig. 4.16). Meanwhile, monitoring sections where the COD concentration exceeded the standard (>10 mg/L) mainly included Shakou, Lifenzha, Xihua Dawangzhuang, Dongsunying, Baogongmiao, Liuzhaicunhou, Niqiu, downstream Linquanreach, Funan, Maqiao, Huangqiao, Yangzhuang, Yigouqiao and Tianchang Chemical Plant, accounting for 64 % of all monitoring sections in secondary tributaries (See Secondary tributaries COD concentration grade (1997–2005), Fig. 4.17). Monitoring sections where the ammonia nitrogen concentration exceeded the standard (>1.5 mg/L) mainly included Lifenzha, Xihua Dawangzhuang, Dongsunying, Baogongmiao, Liuzhaicunhou, Niqiu, downstream Linquan reach, Xuzhuang, Maqiao, Huangqiao, Yangzhuang, Yigouqiao, Tianchang Chemical Plant and Gongnongbingqiao, accounting for 64 % of all monitoring sections in secondary tributaries (See Secondary tributaries ammonia nitrogen grade (1997–2005), Fig. 4.18).

- 2005–2009
  No deteriorated water quality (Grade V or worse) was detected in Dachenzha, Liulidukou, Shakou, Funan, Gongnongbingqiao, Hongshizui, Zhumazha, Xiyingqiao (See Secondary tributaries water quality grade (2005–2009), Fig. 4.19). In contrast, water quality exceeded the standard in all the other 14 monitoring sections for 3 or more years, accounting for 64 % of all monitoring sections of secondary tributaries.

  During this period, the BOD concentration exceeded the standard (>6 mg/L) mainly in Xihua Dawangzhuang, Dongsunying, Lifenzha, downstream Linquan reach, Liuzhaicunhou, Niqiu, Maoqiao, Xuzhuang, Huangqiao, Yangzhuang, Linhuanji and Tianchang Chemical Plant, which accounted for 55 % of all monitoring sections in secondary tributaries (See Secondary tributaries BOD concentration grade (2005–2009), Fig. 4.20). Meanwhile, the COD concentration exceeded the standard (>10 mg/L) mainly in Xihua Dawangzhuang, Dongsunying, Baogongmiao, Liuzhaicunhou, downstream Linquan reach, Maqiao, Linhuanji, Huangqiao and Yangzhuang, which accounted for 41 % of all monitoring sections in secondary tributaries (See Secondary tributaries COD concentration grade (2005–2009), Fig. 4.21). The ammonia nitrogen concentration exceeded the standard (>1.5 mg/L) mainly in the Xihua Dawangzhuang, Lifenzha, downstream Linquan reach, Xuzhuang, Niqiu, Liuzhaicunhou, Dongsunying, Maqiao, Huangqiao, Yangzhuang, Linhuanji, Baogongmiao, Tianchang Chemical Plant and Yigouqiao monitoring sections, accounting for 64 % of all the monitoring sections in secondary tributaries (See Secondary tributaries ammonia nitrogen grade (2005–2009), Fig. 4.22).

  Compared with 1997–2005, the number of monitoring sections in secondary tributaries with deteriorated water quality declined to a certain extent. The concentrations of various pollution indicators decreased, and, especially, the previously higher COD concentration improved significantly. However, in most monitoring sections, BOD and ammonia nitrogen concentrations still remained at a high level and exceeded the standards.
- Water quality grades and indicator concentrations in 1997–2009
  The proportion of monitoring sections where the water quality exceeded the standard (Grade V or worse) in each year fell between 50 and 90 %, and reached as high as 90 % in 2001 (See Proportion of secondary tributaries water quality grade (1997–2009), Fig. 4.23). At the same time, the proportion of the years in which the water quality exceeded the standard (Grade V or worse) in each monitoring section was below 50 % only in Dachenzha, Liulidukou, Funan, Gongnongbing Bridge, Zhumazha Sluice and Xiyingqiao, and above 50 % in the other 16 monitoring sections. Among these monitoring sections, the proportion was 100 % in Shenqiu Lifenzha, downstream Linquan reach and Xuzhuang in the Fenquanhe River, Xihua Dawangzhuang in the Jialuhe River, Niqiu in the Heicihe River, Dongsunying and Liuzhaicunhou in the Guohe River, Yongcheng Maqiao in the Baohe River, as well as Huangqiao and Yangzhuang in the Kuihe River.

  The proportion of monitoring sections where the BOD concentration exceeded the standard (>6 mg/L) in each year fell between 30 and 50 % (See Proportion of secondary tributaries BOD concentration grade (1997–2009), Fig. 4.24). The proportion of the years in which the BOD concentration exceeded the standard was as high as 100 % in some monitoring sections, including downstream Linquan reach and Xuzhuang in the Fenquanhe River, Xihua Dawangzhuang in the Jialuhe River, Dongsunying and Liuzhaicun in the Guohe River, as well as Huangqiao and Yangzhuang in the Kuihe River, which suggested that the BOD concentrations of these monitoring sections exceeded the standard every year.

  The number of monitoring sections where the COD concentration exceeded the standard (>10 mg/L) each year (See Proportion of secondary tributaries COD concentration grade (1997–2009), Fig. 4.25) gradually declined year by year and reached 20 % in 2009. Meanwhile, the proportion of years in which the COD concentration exceeded the relevant standard was above 50 % in the downstream Linquan monitoring section in the Fenquanhe River, the Xihua Dawangzhuang monitoring section in the Jialuhe River, the Dongsunying and Liuzhaicunhou monitoring sections in the Guohe River, and the Maqiao monitoring section in the Baohe River, as well as Huangqiao and Yangzhuang monitoring sections in the Kuihe River. In two of these sections – Maqiao in the Baohe River and Yangzhuang in the Kuihe River – this proportion was as high as 100 %.

  In the same period ammonia nitrogen concentration exceeded the standard (>1.5 mg/L) in very year (See Proportion of secondary tributaries ammonia nitrogen concentration grade (1997–2009), Fig. 4.26) except 2002. In the other years the proportion fluctuated between 25 and 70 %. Meanwhile, the proportion of the years in which the ammonia nitrogen concentration exceeded relevant standard (>1.5 mg/L) was as high as 100 % in the Lifenzha, downstream Linquan reach and Xuzhuang monitoring sections in the Fenquanhe River, the Xihua Dawangzhuang monitoring section in the Jialuhe River, the Niqiu monitoring section in the Heicihe River, the Dongsunying and Liuzhaicunhou monitoring sections in the Guohe River, and the Yangzhuang monitoring section in the Kuihe River. This suggested that the ammonia nitrogen concentrations of these monitoring sections exceeded the standard every year.

Beijing-Hangzhou Canal

- 1997–2005

  The monitoring sections where the water quality exceeded the standard (Grade V or worse) mainly included Nanmatou, Linjiaba and Taierzhuang Bridge in the north section of the canal (See Beijing-Hangzhou Canal water quality grade (1997–2005), Fig. 4.27). However, in the Maling Pumping Station, Baoyingchuanzha and Huaisihekou monitoring sections, the water quality fluctuated between Grade II and Grade IV.

  In this period, Linjiaba was the only monitoring section where the BOD concentration exceeded the standard (>6 mg/L, See Beijing-Hangzhou Canal BOD concentration grade (1997–2005), Fig. 4.28). It was also the only monitoring section where the COD concentration exceeded the standard (>10 mg/L, See Beijing-Hangzhou Canal COD concentration grade (1997–2005), Fig. 4.29). In addition, the ammonia nitrogen concentration exceeded the standard (>1.5 mg/L) only in the Taierzhuang Bridge monitoring section (See Beijing-Hangzhou Canal ammonia nitrogen concentration grade (1997–2005), Fig. 4.30).
- 2005–2009

  Although it exceeded the standard (Grade V or worse) in Taierzhuang Bridge, water quality remained at Grade IV or below Grade IV in all the other monitoring sections (See Beijing-Hangzhou Canal water quality grade (2005–2009), Fig. 4.31).

  In the same period, a higher BOD concentration only appeared in Taierzhuang Bridge monitoring section, and water quality remained at Grade II – Grade III in the Linjiaba, Baoyingchuanzha, Maling Pumping Station, and Huaisihekou monitoring sections (See Beijing-Hangzhou Canal BOD concentration grade (2005–2009), Fig. 4.32). Meanwhile, neither the COD concentration (See Beijing-Hangzhou Canal COD concentration grade (2005–2009), Fig. 4.33) nor the ammonia nitrogen concentration (See Beijing-Hangzhou Canal ammonia nitrogen grade (2005–2009), Fig. 4.34) exceeded the standards in Beijing-Hangzhou Canal.
- Water quality grades and indicator concentrations between 1997 and 2009

  The proportion of the monitoring sections where the water quality exceeded the standard (Grade V or worse) in each year fluctuated between 0 and 50 %, and gradually declined year by year (See Proportion of Beijing-Hangzhou Canal water quality grade (1997–2009), Fig. 4.35). The proportion of the years in which the water quality exceeded the relevant standard was relatively high in Nanmatou, Taierzhuang Bridge and Linjiaba, while the water quality was Grade II and Grade III in Maling Pumping Station, Baoyingchuanzha, Huaisihekou, etc.

  BOD concentrations exceeded the standard (>6 mg/L) only in 1997, 2002, 2006 and 2007, and the proportion of sections with excessive levels of BOD was less than 30 % in all these years (See Proportion of Beijing-Hangzhou Canal BOD concentration grade (1997–2009), Fig. 4.36). The proportion of years in which the BOD concentration exceeded the standard was relatively high in the Taierzhuang Bridge and Linjiaba monitoring sections.

  COD concentrations exceeded the relevant standard (>10 mg/L) only in 1997 and 1999 (See Proportion of Beijing-Hangzhou Canal COD concentration grade (1997–2009), Fig. 4.37). Linjiaba was the only monitoring section where the COD concentration exceeded the standard.

  The ammonia nitrogen concentration exceeded the relevant standard (>1.5 mg/L) only in 1997 (See Proportion of Beijing-Hangzhou Canal ammonia nitrogen concentration grade (1997–2009), Fig. 4.38). Taierzhuang Bridge was the only monitoring section where the ammonia nitrogen concentration exceeded the standard.

Yishusi Water System

- 1997–2005

  Except for the Gangshang and Zhangzhuang monitoring sections where the water quality did not exceed the standard (Grade V or worse), there were instances of water quality exceeding the standard in all the other 17 monitoring sections (See Yishusi water system water quality grade (1997–2005), Fig. 4.39). Some of these monitoring sections showed a deteriorated water quality for 3 or more years, which included Yulou, Malou, Huangzhuang, Shuyuan, Xiyao, Qunleqiao, Shagouqiao, Beiwaihuanqiao, Lijiqiao, Aishan West Bridge, 310 Gongluqiao, Linshu Daxingqiao and Zhangtuanqiao, accounting for 68 % of all monitoring sections in the Yishu-Si River System.

  In this period, the monitoring sections where the BOD concentration exceeded the standard (>6 mg/L) mainly included Yulou, Shuyuan, Huangzhuang, Qunleqiao, Xiyao, Aishan West Bridge, Linshu Daxingqiao, Zhangtuanqiao and Gaofengtou, accounting for 47 % of all monitoring sections in the Yishu-Si River System (See Yishusi water system BOD concentration grade (1997–2005), Fig. 4.41). Meanwhile, the monitoring sections where the COD concentration exceeded the standard (>10 mg/L) were Shuyuan, Qunleqiao, Huangzhuang, Xiyao, Liji Bridge, Yulou, Gaofengtou, etc., accounting for 37 % of all monitoring sections in the Yishu-Si River System (See Yishusi water system COD concentration grade (1997–2005), Fig. 4.42). In addition, there were many monitoring sections where the ammonia nitrogen concentration exceeded the standard (>1.5 mg/L), mainly including Yulou, Huangzhuang, Malou, Qunle Bridge, Liji Bridge, Xiyao, Aishan West Bridge, Linshu Daxing Bridge and Gaofengtou, accounting for 47 % of all monitoring sections in the Yishu-Si River System (See Yishusi water system ammonia nitrogen concentration grade (1997–2005), Fig. 4.43).

- 2005–2009

The number of the monitoring sections where the water quality exceeded the standard (Grade V or worse) declined significantly during this period (See Yishusi water system water quality grade (2005–2009), Fig. 4.44). The deteriorated water quality mainly appeared in the Yulou, Shuyuan, Huangzhuang, Malou, Qunle Bridge, Xiyao, Aishan West Bridge and Dongpianhong monitoring sections, and Yulou, Huangzhuang, Malou, Shuyuan and Qunle Bridge sections showed a deteriorated water quality for 3 or more years, accounting for 26 % of all monitoring sections of the Yishu-Si River System.

In this period, monitoring sections where the BOD concentration exceeded the standard (>6 mg/L) mainly included Yulou, Huangzhuang, Shuyuan, Qunle Bridge, Aishan West Bridge, Gaofengtou and Linshu Daxing Bridge, which accounted for 37 % of all monitoring sections in the Yishu-Si River System (See Yishusi water system BOD concentration grade (2005–2009), Fig. 4.46). The monitoring sections where the COD concentration exceeded the standard (>10 mg/L) mainly included Yulou, Huangzhuang, Shuyuan, Xiyao, Qunle Bridge, Aishan West Bridge, Gaofengtou and Linshu Daxing Bridge, accounting for 42 % of all monitoring sections in the Yishu-Si River System (See Yishusi water system COD concentration grade (2005–2009), Fig. 4.47). Meanwhile, the monitoring sections where the ammonia nitrogen concentration exceeded the standard (>1.5 mg/L) were Yulou, Huangzhuang, Malou, Shuyuan, Qunle Bridge, Aishan West Bridge, Gaofengtou and Linshu Daxing Bridge, accounting for 42 % of all monitoring sections in the Yishu-Si River System (See Yishusi water system ammonia nitrogen grade (2005–2009), Fig. 4.48).

Compared with the situation between 1997 and 2005, the number of the monitoring sections where the water quality exceeded the standard in the Yishu-Si River System declined significantly. However, in certain monitoring sections, the BOD, COD and ammonia nitrogen concentrations still exceeded the standards.

- Water quality grades and indicator concentrations between 1997 and 2009

The proportion of the monitoring sections where the water quality exceeded the standard (Grade V or worse) in each year fluctuated between 15 and 90 %, but it gradually declined year by year after 2003 and reached 21 % in 2009 (See Proportion of Yishusi water system water quality grade (1997–2009), Fig. 4.49). The proportion of the years in which the water quality exceeded the standard fluctuated between 75 and 100 % in the following monitoring sections in the Yishusi – Nansihu region: Yulou in the Zhuzhaoxin River, Huangzhuang in the Guangfuhe River, Shuyuan in the Sihe River, Malou in the Maimahe River, and Qunle Bridge in the Chenghe River. Meanwhile, in the Yishusi-Luomahu Region, the proportion was higher than 50 % only in the Aishan West Bridge monitoring section in the Picang Floodway.

The proportion of monitoring sections where the BOD concentration exceeded the standard (>6 mg/L) in each year was below 40 % in this period, and began to decrease gradually in 2004 (See Proportion of Yishusi water system BOD concentration grade (1997–2009), Fig. 4.50). The proportion of the years in which the BOD concentration exceeded the standard was above 50 % in some monitoring sections including Yulou in the Zhuzhaoxin River, Shuyuan in the Sihe River, Qunle Bridge in the Chenghe River, Aishan West Bridge in the Picang Floodway, and Gaofengtou in the Shuhe River, among which the proportion was as high as 100 % in Yulou monitoring section.

Except for 1997 and 1999, when it was 50 %, the proportion of the monitoring sections where the COD concentration exceeded the standard (>10 mg/L) was below 30 % in almost all years in this period, and gradually decreased year by year (See Proportion of Yishusi water system COD concentration grade (1997–2009), Fig. 4.51). Meanwhile, the proportion of the years in which the COD concentration exceeded the standard was above 50 % only in a few monitoring sections including Shuyuan in the Sihe River, Qunle Bridge in the Chenghe River and Gaofengtou in the Shuhe River, etc.

The proportion of the monitoring sections where the ammonia nitrogen concentration exceeded the standard (>1.5 mg/L) in each year decreased gradually, and no higher ammonia nitrogen concentration was detected in 1999 and 2002 (See Proportion of Yishusi water system ammonia nitrogen concentration grade (1997–2009), Fig. 4.52). The proportion of the years in which the ammonia nitrogen concentration exceeded the standard was above 50 % only in a few monitoring sections such as Yulou in the Zhuzhaoxin River, Qunle Bridge in the Chenghe River, Aishan West Bridge in the Picang Floodway, the Linshu Daxing Bridge in the Xinshuhe River, and Gaofengtou in the Shuhe River.

In terms of the water quality changes in various monitoring sections in the secondary tributaries (See Comparison of other tributaries water quality grade (1997, 2005), Fig. 4.40), the Beijing-Hangzhou Canal and Yishu-Si River System, the rivers with deteriorated or significantly deteriorated water quality mainly distributed in the Yishu-Si River System, as well as the Heicihe River and the Guhe River in Fuyang reach. Since many rivers were seriously polluted in numerous reaches and had a water quality of Grade V or worse in 1997, those rivers whose water quality was classified as unchanged still showed a high level of pollution in 2005; for those river reaches whose water quality had improved, the dominant water quality was still at Grade IV. These monitoring results indicated that the water pollution in the tributaries in the Huai River Basin had declined to a certain extent in some reaches, but the problem had not been fundamentally resolved, and serious water pollution still existed in some places. Although an increase was observed in the number of river reaches in which the water quality had improved or significantly improved (See Comparison of other tributaries water quality grade (2005, 2009), Fig. 4.45), together with a decrease in the number of river reaches in which the water quality had deteriorated or significantly deteriorated, the water quality in most river reaches remained unchanged. Therefore, even though water quality had improved to a certain extent in some reaches in this period, overall pollution was still very serious in the tributaries in the Huai River Basin.

Lakes

The water in Hongze Lake was seriously polluted for many years, with water quality of Grade V or worse (see Water quality grade (1983–2009), Fig. 5.1). The main pollution indicators were total nitrogen and total phosphorus. In 1993, 2002 and 2003, there was a short period of improvement (Grade IV), but the water quality quickly deteriorated again to below Grade V. Nansi Lake also experienced serious pollution, with water quality with worse than Grade V in 2005. The main pollution indicators were total nitrogen, total phosphorus and BOD. Since 2006, water quality has improved somewhat and reached Grade IV in 2008 and in 2009.

Spatiotemporal Variation of Water Environment in Huai River Basin

Analysis of the data on changes in water quality in the monitoring sections of the Huai River from 1982 to 2009, shows that the water quality has the following general characteristics:

(a) Time: there has been fluctuation in levels of water quality over time: in some years there was serious water pollution and in some years water quality improved;
(b) Spatial distribution: there was variation across branches, with serious water pollution in tributaries and lakes over many years;
(c) Pollution indicators: indicators of serious water pollution in the main channel and tributaries are ammonia/non-ionic ammonia, 5-day biochemical oxygen demand (BOD), chemical oxygen demand (COD).

In each of the seven regions, which were divided according to their terrain and the direction of rivers, the proportions of sections with water quality ranked Grade II, Grade III, Grade IV, and Grade V and worse than V were calculated for each year from 1997 to 2009 (see Water quality in different regions (1997–2009), Fig. 6.1). Water quality in the western hilly region was the best; and water quality in the middle reaches of the western plains and Nansi Lake was the worst. In terms of temporal variation, the proportion of regions with water quality ranked Class V or worse than V has declined.

From 1997 to 2009, regions in which water pollution was most serious were mainly on the north bank of the Huai River (see Frequency of water pollution (1997–2009, 2001–2009, and 2005–2009), Figs. 6.2, 6.3, and 6.4). Water pollution was particularly serious near the Hong-Fenquan-Ying-Guo River (central-west plain), Kui River (central-east plain), as well as the Zhuzhaoxin-Si River (Nansi water system) regions. Over time, however, areas with a high frequency of serious pollution gradually decreased. Because the FWPs of BOD (see Frequency of BOD pollution (1997–2009 and 2005–2009), Figs. 6.5 and 6.6) and ammonia nitrogen (see Frequency of ammonia nitrogen pollution (1997–2009 and 2005–2009), Figs. 6.9 and 6.10) were very high and their distribution was consistent with that of water quality of Grade V and worse, we judged that the most important pollution indicators of water quality for the period were BOD and ammonia nitrogen.

Variation in Mortality from Digestive Cancers in the Huai River Basin

According to the 1973–1975 national cause of death survey, cancer mortality rates for both males and females in most counties/districts along the Huai river basin (with the exception of eastern Jiangsu province) were lower than the national level (the standardized mortality rate was 66.92/100,000), and were regarded as low-incidence areas for cancer. (see Age-standardised males mortality rate for liver cancer (1973–1975), Fig. 7.1 and Age-standardised females mortality rate for liver cancer (1973–1975), Fig. 7.2) Among the 14 surveillance counties/districts, some counties of the eastern lower reaches of the Huai River, such as Xuyi County, Jinhu County and Sheyang County were "high cancer incidence areas", where the cancer mortality rates were 1.5 times to 2 times higher than the national level; Half of the counties belonged to "lower cancer incidence areas", where cancer mortality rates were 60–80 % of the national level; And others were similar to the national level (with cancer mortality rates 80–120 % of the national level). And in these areas in 2004–2006, not only did places which originally had high rates of mortality from cancer continue to have death rates above the national average, but some counties and cities which originally had low cancer mortality rates now had rates that exceeded the national average by more than 20 %, with some places exceeding it by as much as 60 % (see Age-standardised mortality rate of digestive cancer (2004–2006), Fig. 7.3). In terms of the scale of change, from 1973 to 2006, the national age-adjusted cancer mortality rate increased by 20.16 %. With the exception of counties that originally had a high mortality rate, which had normal increases of 16–18 %, nearly all other areas saw an increase in mortality of more than 20 %, and in some places it was once or twice as high as that (see Change in rates of age-adjusted cancer mortality, 1973–2006, Fig. 7.4).

The increase in cancer mortality rates was associated with the change in mortality rates of some kinds of digestive cancers. This atlas describes in detail the change in mortality rates for liver cancer, gastric cancer and esophagus cancer over the past 30 years.

1. **Liver cancer**: In 1973–1975, liver cancer mortality rates in Jinghu, Xuyi and Sheyang in the lower reaches of the Huai River were higher than the national level. However, liver cancer mortality in Shenqiu and Yingdong counties on the Shaying River, in Fugou and Mengchen counties on the Guo River, in Yongqiao and Lingbi counties on the Kui River, in Xiping county on the Hong River (the West Plain area of the upper-middle reaches), and in Wenshang and Juye counties in the basin of the Yi, Shu and Si Rivers were about 50 % of the national level in most cases and equal to or below the national average in a few others. (see Age-standardised males mortality rate for liver cancer (1973–1975), Fig. 7.5 and Age-standardised females mortality rate for liver cancer (1973–1975), Fig. 7.6) By 2004–2006, the liver cancer mortality rate among the population of these regions was higher than the national average in almost all cases. In particular, liver cancer mortality rates in the population of the Shaying, Kui, and Guo river basins was 1.45–1.86 times higher than the national average. (see Age-standardised males mortality rate for liver cancer (2004–2006), Fig. 7.7 and Age-standardised females mortality rate for liver cancer (2004–2006), Fig. 7.8) More disturbingly, the increase in liver cancer mortality among the population in these originally low-incidence areas was more than twice the national average rate of increase in the disease. In Shenqiu it was as much as 5.43 times (see Change in rates of age-adjusted liver cancer mortality, 1973–2006, Fig. 7.9).
2. **Gastric cancer**: In 1973–1975, gastric cancer mortality rates in Jinghu, Xuyi, Sheyang and Luoshan counties were higher than the national average; other areas had a low incidence of gastric cancer, with mortality rates only 40–80 % of the national average. (see Age-standardised males mortality rate for liver cancer (1973–1975), Fig. 7.10 and Age-standardised females mortality rate for liver cancer (1973–1975), Fig. 7.11) In 2004–2006, mortality from gastric cancer in the previously high mortality areas of Jinhu and Sheyang was still above the national average, but the speed of decline was quicker than the national average, and the gap between the two had shrunk. In the majority of areas that were originally low risk for gastric cancer, mortality rates showed a trend towards increase and were on average higher than the national rate. (see Age-standardised males mortality rate for liver cancer (2004–2006), Fig. 7.12 and Age-standardised females mortality rate for liver cancer (2004–2006), Fig. 7.13) In particular, mortality from gastric cancer in Shenqiu was 2.56 times higher than the mortality rate in the 1975 (see Change in rates of age-adjusted stomach cancer mortality, 1973–2006, Fig. 7.14).
3. **Esophageal cancer**: Mortality from esophagus cancer in this area has been going down over the past 30 years. In 1973–1975, mortality rates for esophageal cancer in Jinhu, Xuyi, Sheyang, and Yindong counties were higher than the national level; the mortality rates in the other counties were the same as or slightly lower than the national rate. (see Age-standardised males mortality rate for liver cancer (1973–1975), Fig. 7.15 and Age-standardised females mortality rate for liver cancer (1973–1975), Fig. 7.16) During 2004–2006, mortality rates for esophageal cancer in most areas of the 14 surveillance counties in the Huai River Basin were higher than the national average, including Shenqiu County, Shou County, Yingdong County, Lingbi County and Mengcheng County, as well as Wenshang and Juye in the Yi, Shu and Si River Basin (see Age-standardised males mortality rate for liver cancer (2004–2006), Fig. 7.17 and Age-standardised females mortality rate for liver cancer (2004–2006), Fig. 7.18). In general, the age-standardised death rates for esophageal cancer at the national level and in most areas were in decline. (see Change in rates of age-adjusted esophageal cancer mortality, 1973–2006, Fig. 7.19)
4. **Comparison of mortality rates of several cancers with the national level**: Table 1.3 shows mortality rates for the three cancers (liver cancer, gastric cancer and esophagus cancer) in 1973–1975 and 2004–2005, and comparison with the national levels of mortality from the three cancers. The national standardized mortality rate for liver cancer has increased by 62 % over this 30 year period, while mortality rates for gastric cancer and esophagus cancer have been going down, decreasing by 9 % and 43 %, respectively.

The results of the 1973 survey show that some counties of the eastern lower reaches of the Huai River, such as Jinghua County, Sheyang County and Xuyi County, were "high cancer incidence areas". Over the past 30 years, the rate of increase in deaths from liver cancer has been lower than the national average, and the rate of decrease in deaths from gastric and esophageal cancer has been faster than the national average. However, some counties in the western plain of the middle reaches of the Huai River, such as Shenqiu, Yindong and Yongqiao, were "low cancer incidence areas" in 1975, yet the increasing rate of liver cancer mortality has exceeded the national average by several times, and mortality from gastric cancer and esophagus cancer, which have decreased nationally, have been increasing in these areas.

Table 1.3 Analysis of levels and trends in deaths from liver, gastric, and esophageal cancer in 14 counties of the Huai River Basin from 1973–1975 and 2004–2006

Counties	Water sheds	Liver Cancer					Gastric Cancer					Esophageal cancer				
		Annual death rate ($1/10^5$) (1973-1975)	Annual death rate ($1/10^5$) (2004-2006)	Compared to nation (1973-1975)	Compared to nation (2004-2006)	Speed of change	Annual death rate ($1/10^5$) (1973-1975)	Annual death rate $1/10^5$) (2004-2006)	Compared to nation (1973-1975)	Compared to nation (2004-2006)	Speed of change	Annual death rate ($1/10^5$) (1973-1975)	Annual death rate ($1/10^5$) (2004-2006)	Compared to nation (1973-1975)	Compared to nation (2004-2006)	Speed of change
Nation		14.45	23.4	1.00	1.00	0.62	23.67	21.57	1.00	1.00	-0.09	23	13.21	1.00	1.00	-0.43
Luo shan	Main stream	14.45	25.61	1.00	1.09	0.77	26.72	37.78	1.13	1.75	0.41	30.99	10.72	1.35	0.81	-0.65
Shou		9.91	24.58	0.69	1.05	1.48	18.6	33.25	0.79	1.54	0.79	21.05	21.33	0.92	1.61	0.01
Xuyi		17.48	26.6	1.21	1.14	0.52	32.02	34.3	1.35	1.59	0.07	58.43	44.69	2.54	3.38	-0.24
Jinhu		23.79	23.28	1.65	0.99	-0.02	57.91	44.88	2.45	2.08	-0.23	77.81	50.87	3.38	3.85	-0.35
Sheyang		31.6	37.15	2.19	1.59	0.18	62.47	46.44	2.64	2.15	-0.26	79.15	33.82	3.44	2.56	-0.57
Xiping	Hong River	19.61	25.24	1.36	1.08	0.29	20.56	20.15	0.87	0.93	-0.02	42.69	18.8	1.86	1.42	-0.56
Shenqiu	Shaying River	6.76	43.44	0.47	1.86	5.43	10.01	35.66	0.42	1.65	2.56	25.24	29.02	1.10	2.20	0.15
Yingdong		15.75	36.94	1.09	1.58	1.35	13.54	35.15	0.57	1.63	1.60	35.33	34.7	1.54	2.63	-0.02
Fugou	Guo River	10.68	33.98	0.74	1.45	2.18	9.44	16.69	0.40	0.77	0.77	15.36	10.26	0.67	0.78	-0.33
Mengcheng		7.36	34.56	0.51	1.48	3.70	9.58	19.21	0.40	0.89	1.01	19.9	20.14	0.87	1.52	0.01
Yongqiao	Kui River	10.92	24.54	0.76	1.05	1.25	13.32	23.31	0.56	1.08	0.75	15.53	12.54	0.68	0.95	-0.19
Linbi		11.17	38.47	0.77	1.64	2.44	12.12	26.75	0.51	1.24	1.21	22.09	16.81	0.96	1.27	-0.24
Wenshang	Yishusi Water System	5.98	22.36	0.41	0.96	2.74	11.89	28.12	0.50	1.30	1.37	45.94	42.79	2.00	3.24	-0.07
Juye		9.57	19.67	0.66	0.84	1.06	12.87	21.06	0.54	0.98	0.64	24.5	16.43	1.07	1.24	-0.33

Conclusion

Through integrated analysis of surface water monitoring data and cause of death survey/surveillance data for the Huai River, this Atlas has displayed the history of surface water pollution and the profile of mortality from digestive tract cancers over the past 30 years.

Analysis of water quality monitoring data from the 86 state-controlled sections of the Huai River from 1982 to 2009 shows that the main stream, tributaries and lakes of the river basin have experienced varying degrees of water pollution. The most serious pollution occurred in 1996–1997, 2001–2002 and 2004–2005. After 2005, the situation improved. The main indicators of pollution in the Huai River Basin were ammonia/non-ionic ammonia, BOD, and COD.

There were no serious instances of water pollution in the main stream of Huai River in 2007–2009, i.e. there were no instances of water quality of Grade V or worse. Although the number of sections with water quality of Grade V or worse in various tributaries had decreased, the problem persisted and was mainly concentrated in the Fenquan River, Shaying River, and Guo River in the central-west plain and the Kui River in the central-east plain of the Huai River Basin. Water pollution in the region's lakes was still very serious. The pollution indicators for the main channel and tributaries are somewhat different, and include total nitrogen and total phosphorus in addition to BOD.

A comparison of cause of death data for the Huai River Basin in 1973–1975 and 2004–2005 shows that mortality rates for digestive cancers, especially liver cancer and gastric cancer, rapidly shifted from low to high in the areas of Shenqiu and Yingdong on the Shaying River, Fugou and Mengcheng in the Guo River Basin, Yongqiao and Lingbi in the Kui River Basin, and Wenshang and Juye on the Yishusi water system. The mortality rate increased several times faster than the national average. These areas were precisely the ones with the most serious and long-term pollution contamination, including the tributary areas of the Hong, Shaying, Guo and Kui rivers. There is a strong spatial consistency between the polluted areas and those in which the increase in digestive cancers was the largest.

The causes of cancer are complex and it is hard to verify the correlation between pollution and cancer incidence just through analysis of their temporal and spatial consistency. This report is based on the program, *the correlation between cancer and the risk factors along the Huai River Basin*. In 2005, we compared the cancer incidence, mortality rate and the carcinogenic contaminants between study areas (adjacent to Huan River or its tributaries) and control areas (far from Huai River and its tributaries); In addition, we investigated the type, size, numbers and distribution for polluted manufactories (including the current and closed factories since 1985), as well as product procedure, potential poison material and exhaust emissions, in the above two areas.

In addition, we also conducted a household survey for risk factors correlated with lung cancer, esophagus cancer, gastric cancer and liver cancer. The survey included the liver environment (such as fuel for cooking), individual behavior (e.g.: smoking, alcohol), dietary (e.g.: smoked or mildew food intake), biotic components (e.g.: hepatitis B infection and helicobacter pylori infection) and family members' health status (e.g.: cancer history for family members). The survey results indicated that with the exception of the cancer history for the immediate family members, other risk factors showed no significant difference between study areas and control areas. Therefore these risk factors cannot explain why the cancer incidence was higher in the upstream of Huai River Basin. Cancer history of immediate family members is an indicator which can reflect both genetic factors and environmental factors. So considering the geographic positions of the study areas and control areas, the proportion of immediate family members with a history of cancer was consistent with the cancer incidence among the population, which could further verify the correlation between high cancer incidence and environmental factors.

Therefore this atlas shows a high correlation between the seriously polluted areas and areas with high mortality from cancer based on temporal and spatial analysis, and it makes a good case for research on the relationship between water pollution and cancer incidence. However, this level of evidence can only show that there is a correlation between the incidence of cancer and environmental quality and cannot demonstrate a cause-effect relationship.

Although pollution in the Huai River Basin was basically brought under control after 2005, relatively serious water pollution problems persist in some parts of the tributaries of the Huai River. People in some areas are still faced with a higher risk of contracting and dying from cancer. Considering that there is a latency period in the health effects of environmental pollution, it can be expected that in the next 10 years, the Huai River Basin, especially the central and north-central regions where the ammonia / non-ionic ammonia, BOD, COD and other water quality indicators show high levels of pollution, will continue to face a serious situation in terms of cancer prevention and control.

There is a need for government at all levels and relevant departments to strengthen their cooperation and step up pollution control in the Huai River Basin. The emission of pollutants – in particular organic pollutants – needs to be reduced, and ecological conditions improved, especially the water environment. More attention must also be paid to cancer prevention and control in order to reduce the risk to the health of the public in the area.

Appendix

Variation in water quality in monitoring sections of the main stream

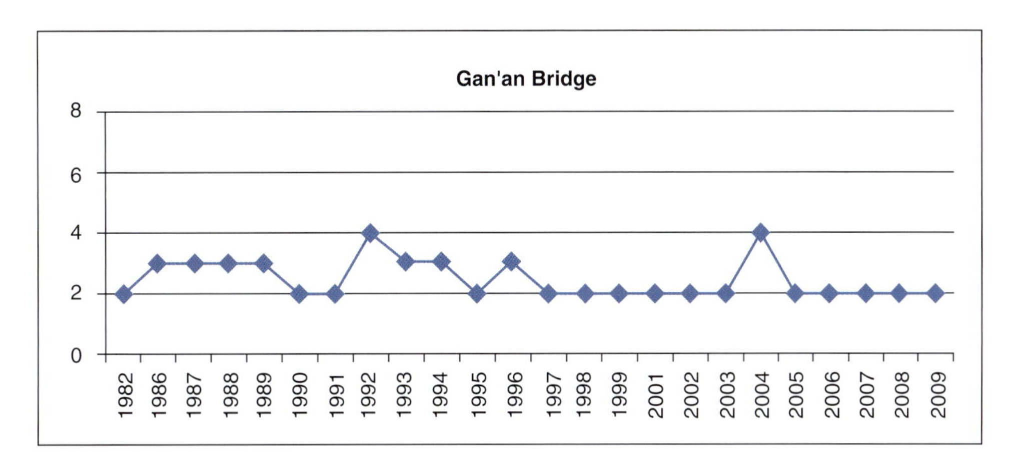

Water quality at Gan'an Bridge in Xinyang City changed between Grade II and Grade III from 1982–2009. Except in 1992 and 2004 when water quality was Grade IV, it was Grade II and Grade III in 92 % of years monitored, which indicates that water quality in the waterhead area is generally good

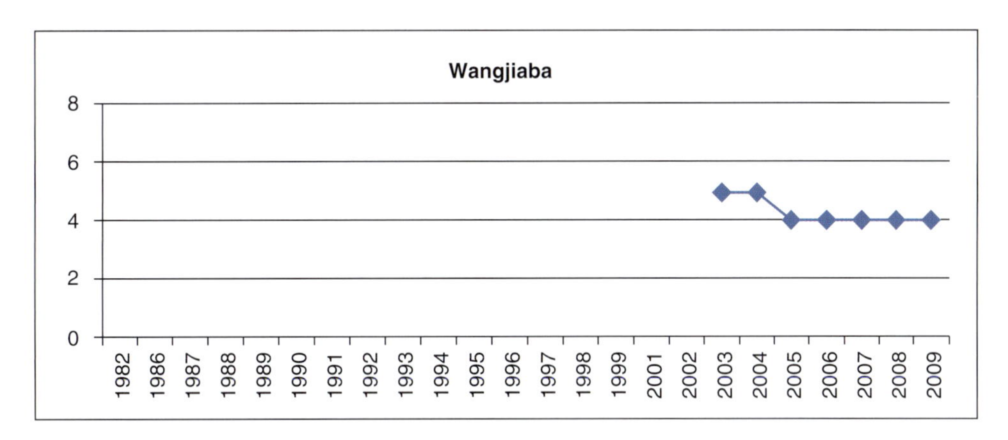

The monitoring section at Wangjiaba in Fuyang City was not established until 2003 so there are only 7 years of monitoring results. Water quality in this section was Grade V in 2003–2004 and Grade IV in 2005–2009

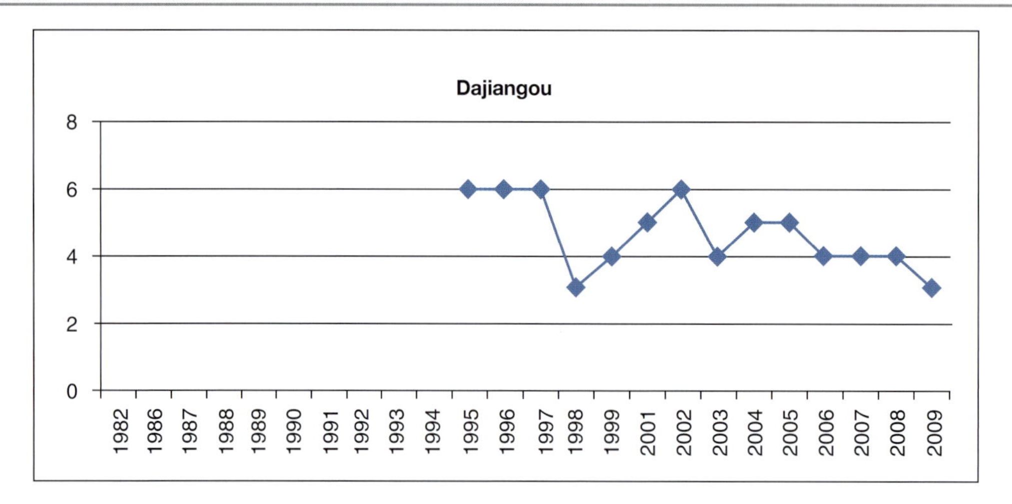

Water quality in this section was Grade V or worse from 1995–1997, although it was Grade III in 1998 and Grade IV 2003. It began to improve from 2006 on and it was Grade III in 2009. Overall, the amount of years with water quality of Grade V or worse accounted for 50 % of all years for which there is monitoring data

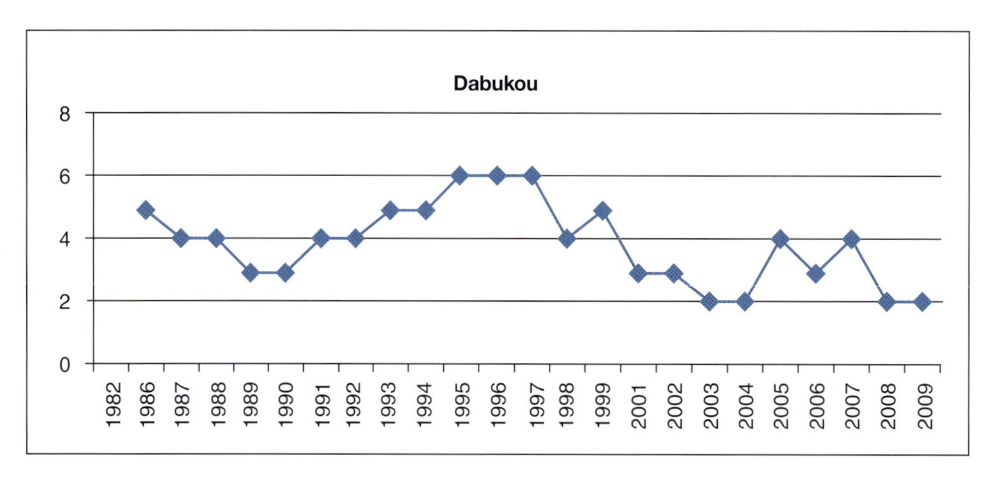

Although water quality of Dabukou section in Xi County improved in 1986, it started deteriorating from 1991 and it was Grade V or worse from 1995 to 1997. There was a second improvement in water quality in this section during 1998–2009 when it remained at Grade II

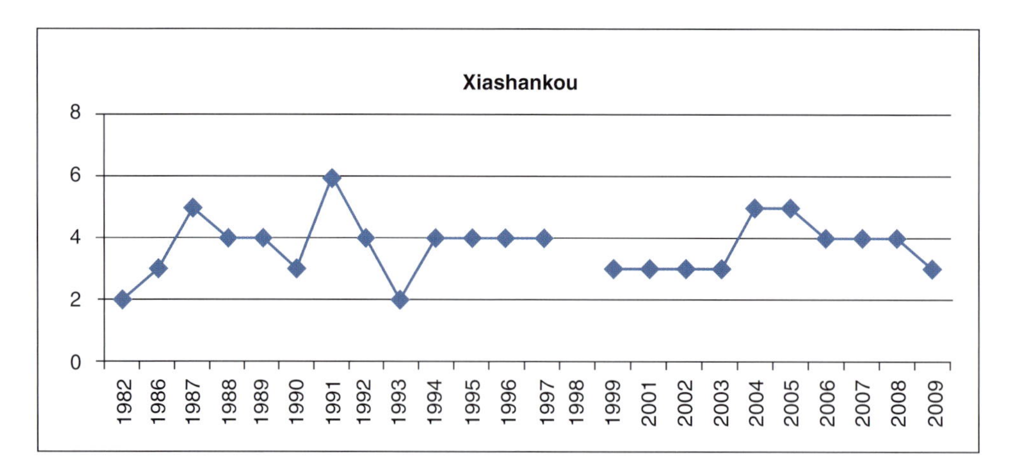

Water quality in the Xiashankou section in Huainan City varied from Grade II to Grade VI or worse from 1982 to 2009. The years with Grade V or worse were 1987, 1991, and 2004–2005, which accounted for about 17 % of all for which there is monitoring data

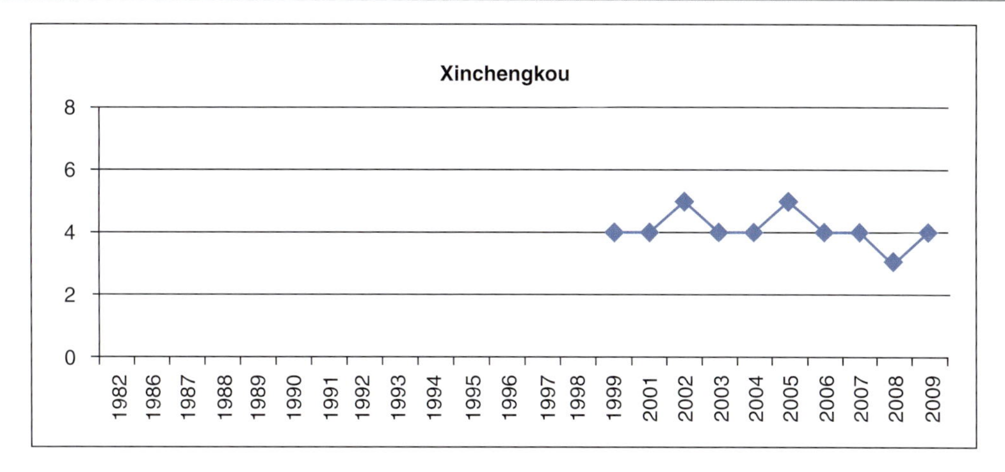

The worst water quality (Grade V) was recorded in 2002 and 2005, accounting for one-fifth of all years. About 70 % of all years for which there is monitoring data reported water quality of Grade IV

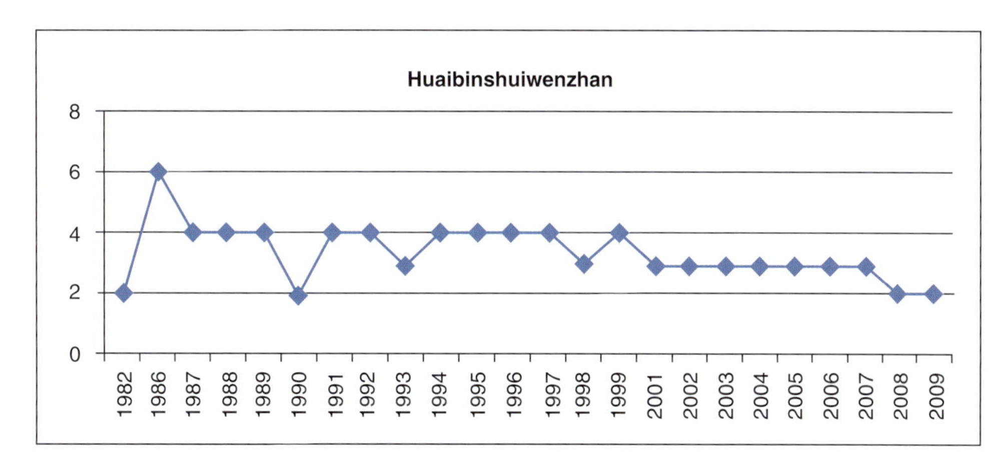

Grade V or worse water quality occurred only once in the Huaibin Section in Xinyang City in 1986. After this, the water quality of this section remained at Grade II-III-IV

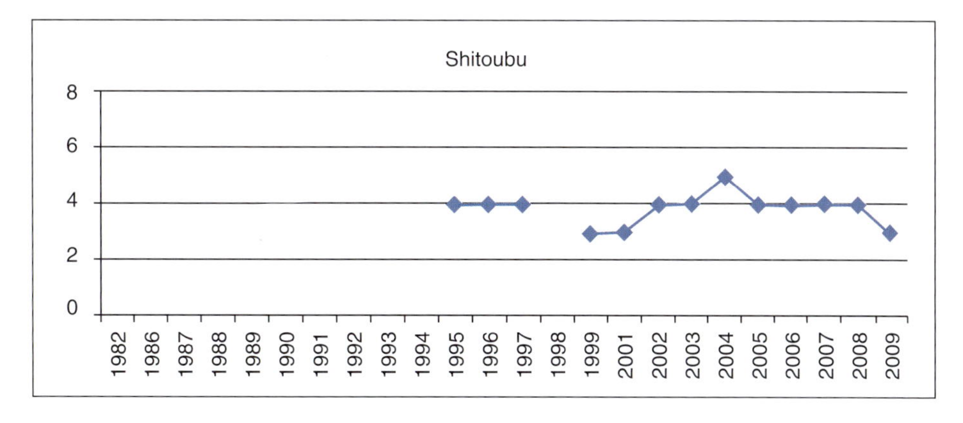

For the Shitoubu Section in Huainan City, Grade IV water quality was reported for most years, accounting for about 69 % of all years. Grade V appeared only in 2004 and Grade III was recorded for the other years (1999, 2001, and 2009)

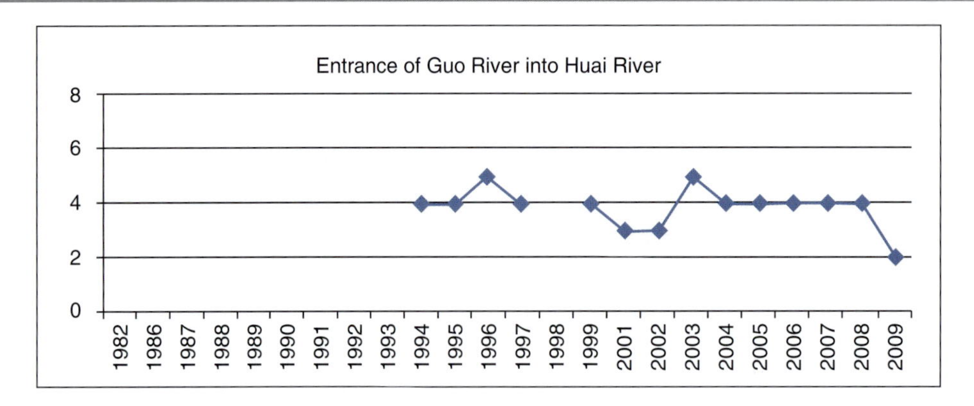

Of the 14 years with monitoring results for the section at the entrance of the Guo River into Huai River, years with Grade IV water quality accounted for about 64 %. Grade V was observed only in 1996 and 2003

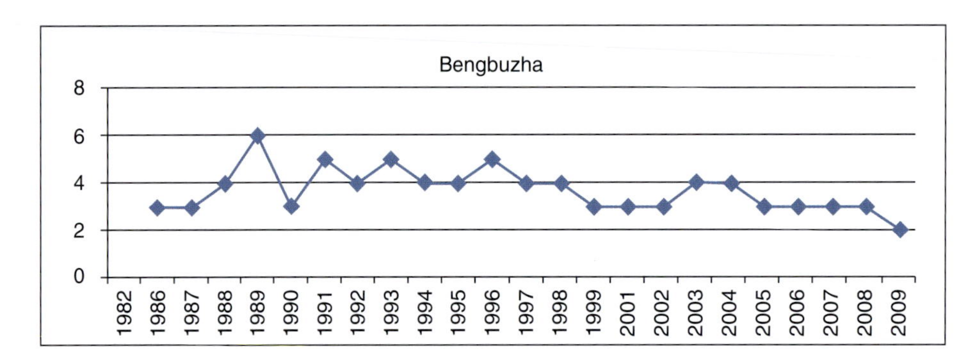

The Bengbuzha Section had water quality of Grade II and Grade III (48 %), Grade IV (35 %) and Grade V or worse (17 %) in various years, which shows that this segment of the mainstream of Huai River was once seriously polluted but has now shown some improvement

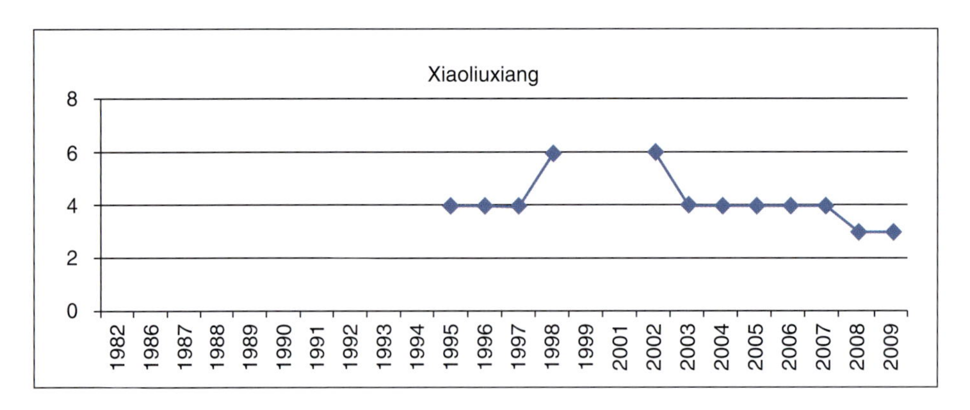

Water quality in the Xiaoliuxiang Section in Chuzhou City was Grade IV from 1995–1997 and 2003–2007, or about 67 % of the time. Grade V or worse water quality was recorded in 1998 and 2002, and the best quality was Grade III during 2008–2009

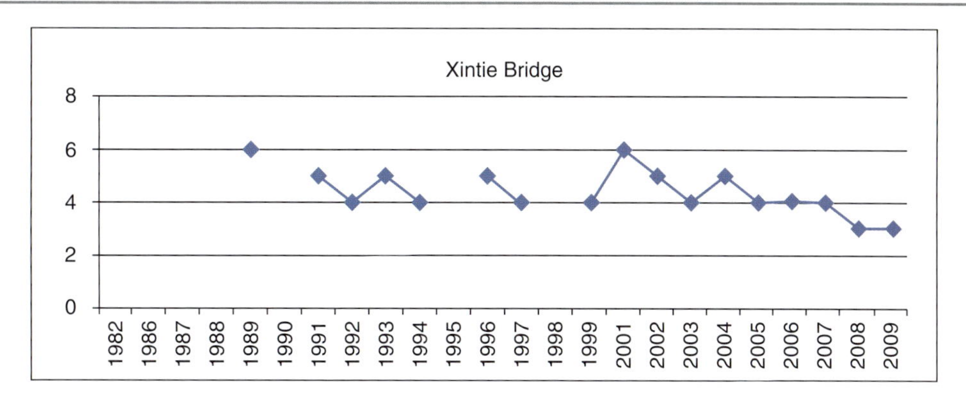

The Xintie Bridge Section had water quality of Grade IV 47 % of the time and of Grade V or worse 41 % of the time (1989, 1991, 1993, 1996, 2001, 2002, and 2004), which shows that this segment of the Huai River was seriously polluted

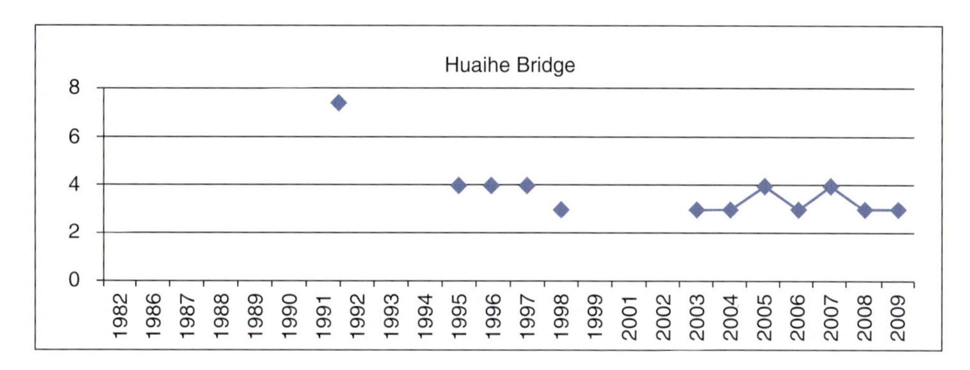

Water quality for the Huaihe Bridge Section in Xuyi County, varied between Grade III and Grade IV

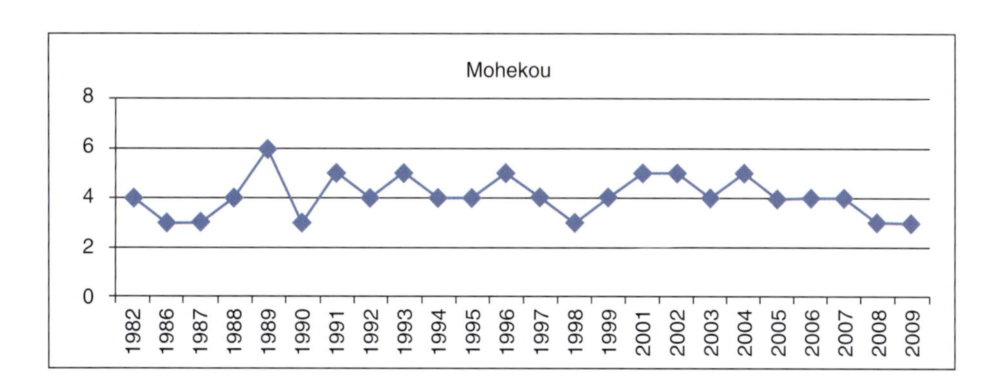

Water quality of Grade IV or better was mostly reported for the Mohekou Section and Grade V or worse water quality mainly appeared in 1989, 1991, 1993, 1996, 2001, 2002 and 2004

Tables

Table 1.4 Variation in water quality grades for monitoring sections of the main stream of the Huai River, 1982–2009

Number of sections / Year	A	B	C	D	E	F	G	H	I	J	K	L	M	N	Annual proportion of sections with Grade V or worse
	A, B...N represent (upper, middle and lower reaches): Chang Tai Guan Gan Bridge, Dabukou, Huaibin, Wangjiaba, Xiashankou, Shitoubu Dajiangou, Xinchengkou, Entrance of Guo River into Huai River, Bengbuzha, Xintie Bridge, Mohekou, Xiaoliuxiang, Xuyi Huai River Bridge														
Number of years	24	23	24	7	23	13	14	10	14	23	17	24	12	11	
1982	2		2	Missing data	2	Missing data		Missing data	Missing data			4	Missing data	Missing data	
1986	3	5	6		3					3		3			
1987	3	4	4		5					3		3			
1988	3	4	4		4					4		4			
1989	3	3	4		4					6	6	6			
1990	2	3	2		3					3		3			
1991	2	4	4		6					5	5	5			
1992	4	4	4		4					4	4	4			
1993	3	5	3		2					5	5	5			
1994	3	5	4		4				4	4	4	4			
1995	2	6	4		4	4	6		4	4		4	4	4	18%
1996	3	6	4		4	4	6		5	5	5	5	4	4	50%
1997	2	6	4		4	4	6		4	4	4	4	4	4	17%
1998	2	4	3				3			4		3	6	3	13%
1999	2	5	4		3	3	4	4	4	3	4	4	Missing data	Missing data	9%
2001	2	3	3		3	3	5	4	3	3	6	5			27%
2002	2	3	3		3	4	6	5	3	3	5	5	6		42%
2003	2	2	3	5	3	4	4	4	5	4	4	4	4	3	14%
2004	4	2	3	5	5	5	5	4	4	4	5	5	4	3	43%
2005	2	4	3	4	5	4	5	5	4	3	4	4	4	4	21%
2006	2	3	3	4	4	4	4	4	4	3	4	4	4	3	0
2007	2	4	3	4	4	4	4	4	4	3	4	4	4	4	0
2008	2	2	2	4	4	4	4	3	4	3	3	3	3	3	0
2009	2	2	2	4	3	3	3	4	2	2	3	3	3	3	0
Proportion of water quality Grades for each section: II - III	92%	39%	55%			23%	14%	10%	21%	48%	12%	25%	17%	55%	
Proportion of water quality Grades for each section: IV	8%	30%	42%		42%	69%	36%	70%	64%	35%	47%	46%	67%	45%	
Proportion of water quality Grades for each section: V or worse		31%	3%			8%	50%	20%	15%	17%	41%	29%	16%	/	

Table 1.5 Variation in water quality grades for monitoring sections of the main tributaries of the Huai River, 1997–2009

Sections	River	Regions	1997	1998	1999	2001	2002	2003	2004	2005	2006	2007	2008	2009	Proportion of water quality grades for each section		
															II-III	II-III	II-III
Pipashan Bridge	Shi	Xinyang	6	6	5	5	4	3	4	4	4	4	3	3	25%	42%	33%
Huanchuan	Huang	Xinyang	4	2	2	5	2	2	4	4	4	3	2	2	58%	33%	8%
Bantai	Hong	Zhumadian	0	6	6	5	6	6	5	5	4	4	4	4	0%	36%	64%
Taolao		Fuyang	0	6	6	0	6	6	6	5	4	4	4	4	0%	40%	60%
Jiangji	Shiguan	Xinyang	3	2	2	2	2	2	2	4	2	2	2	2	92%	8%	0%
Zhidian	Ying	Zhoukou	6	6	6	6	6	6	6	6	6	6	6	6	0%	0%	100%
Yingshang Lower		Fuyang	5	4	6	6	6	6	6	6	6	6	5	4	0%	17%	83%
Jieshou		Fuyang	4	6	6	0	6	6	6	6	6	6	6	5	0%	9%	91%
Huaidian		Zhoukou	0	0	0	0	0	6	6	6	6	6	6	6	0%	0%	100%
Xinandu	Pi	Liu'an	5	4	4	6	5	4	5	6	3	3	3	2	33%	25%	42%
Lixin	Xifei	Bozhou	3	3	4	5	4	4	4	4	3	3	3	4	42%	50%	8%
WuliZha	Dongfei	Liu'an	4	6	6	6	3	3	6	3	3	3	3	3	58%	8%	33%
Fu Bridge	Guo	Zhoukou	6	6	6	6	6	6	5	5	6	6	6	5	0%	0%	100%
Bozhou		Bozhou	6	4	6	0	6	6	6	6	6	6	6	6	0%	9%	91%
Mengcheng		Bozhou	5	4	6	6	6	6	6	6	6	6	6	5	0%	8%	92%
Huangkou	Hui	Shangqiu	0	0	0	0	6	6	4	5	5	4	4	4	0%	50%	50%
Guzhen		Bengbu	0	2	6	3	5	5	4	4	4	4	3	4	27%	45%	27%
Gonglu Bridge	Chi	Chuzhou	4	4	6	6	6	4	6	6	6	6	4	4	0%	42%	58%
Zhang Bridge	Tuo	Shangqiu	0	6	6	0	0	3	4	4	4	4	4	4	11%	67%	22%
Xiaowang Bridge		Huaibei	0	0	0	0	0	4	4	4	4	4	4	4	0%	100%	0%
Bali Bridge	Xinsui	Suzhou	0	0	6	0	6	6	6	6	6	6	5	6	0%	0%	100%
Daqu		Suqian	4	4	5	3	4	6	4	4	4	5	6	6	8%	50%	42%
Gonglu Bridge	Xinbian	Suzhou	0	0	6	0	6	6	6	6	5	4	4	4	0%	33%	67%
Qubei	Irrigation channel	Huaian	3	2	2	4	4	4	3	3	3	2	3	3	75%	25%	0%
Suzui	Drainage channel	Huaian	4	4	3	6	6	5	6	6	6	5	5	3	17%	17%	67%
Annual proportion of sections with Grade V or worse			44%	40%	73%	75%	68%	60%	60%	60%	48%	44%	40%	32%			

Table 1.6 Variation in water quality grades for monitoring sections of the secondary tributaries of the Huai River

Sections	River	Regions	1997	1998	1999	2001	2002	2003	2004	2005	2006	2007	2008	2009	Proportion of water quality grades for each section		
															II-III	II-III	II-III
Tianchang	Baita	Chuzhou	5	3	6	6	5	4	6	6	5	5	4	5	8%	17%	75%
Yigou Bridge	Beichenzi	Gaoyou	4	4	4	6	6	6	6	6	4	5	4	5	0%	42%	58%
Baogongmiao		Shangqiu	6	6	6	0	5	4	4	5	5	4	5	5	0%	27%	73%
Liuzhaicun	Dasha	Bozhou	0	0	0	0	0	6	6	6	6	6	6	6	0%	0%	100%
Linquan		Fuyang	0	6	6	6	6	6	6	6	6	6	6	6	0%	0%	100%
Lifen		Zhoukou	6	6	6	0	0	6	6	6	6	6	6	6	0%	0%	100%
Xuzhuang	Fenquan	Fuyang	0	0	0	0	0	6	5	6	6	6	6	5	0%	0%	100%
Gongnongbing Bridge	Feng	Liu'an	5	0	4	6	3	4	3	4	3	4	3	3	45%	36%	18%
Funan	Gu	Funan	0	6	5	6	6	4	4	4	4	4	4	4	0%	64%	36%
Niqiu	Heici	Fuyang	0	5	6	0	6	6	6	6	6	6	6	6	0%	0%	100%
Linhuanji		Huaibei	0	3	6	0	6	4	4	4	5	5	4	5	10%	40%	50%
Ma Bridge	Hui	Shangqiu	6	6	6	6	5	5	6	6	6	6	6	6	0%	0%	100%
Dawangzhuang	Jialu	Zhoukou	5	6	6	6	6	6	6	6	6	6	6	6	0%	0%	100%
Huang Bridge	Kui	Xuzhou	6	6	6	0	6	6	6	6	6	5	6	6	0%	0%	100%
Shakou	Ru	Zhumadian	4	6	5	6	5	5	5	4	4	4	4	4	0%	50%	50%
Liulidukou	Sha	Luohe	4	4	4	6	4	6	3	3	4	3	2	2	42%	42%	17%
Hongshizui	Shi	Liu'an	2	6	2	0	2	2	2	2	3	2	2	2	91%	0%	9%
Xiying Bridge	Tongyang canal	Nantong	0	4	4	6	5	5	3	3	4	3	4	3	36%	36%	27%
Dongsunying	Guo	Zhoukou	6	6	6	0	6	6	6	6	6	6	6	6	0%	0%	100%
Yangzhuang	Xinsui	Suzhou	6	0	0	0	0	6	6	6	6	6	6	6	0%	0%	100%
Zhumazha	Yan	Huai'an	4	4	2	4	4	4	3	4	4	4	3	3	33%	67%	0%
Dachenzha	Ying	Pingdingshan	3	0	0	0	0	2	2	2	2	2	2	2	100%	0%	0%
Annual proportionof sections with Grade V or worse			60%	65%	67%	91%	76%	64%	59%	59%	59%	59%	50%	64%			

Table 1.7 Variation in water quality grades for monitoring sections of Beijing-Hangzhou Canal

Sections	Regions	1997	1998	1999	2001	2002	2003	2004	2005	2006	2007	2008	2009	Proportion of water quality grades for each section		
														II -III	II -III	II -III
Baoyingchuanzha	Yangzhou	0	3	4	3	3	3	3	4	3	3	3	3	82%	18%	0%
Huaisihekou	Yangzhou	4	2	2	4	3	3	3	3	2	2	2	2	83%	17%	0%
Linjiaba	Xuzhou	5	6	5	4	5	6	3	3	3	3	3	3	50%	8%	42%
Malin	Suqian	4	3	3	4	4	3	3	3	3	3	3	3	75%	25%	0%
Nanmatou	Jining	0	6	6	0	0	5	4	4	4	4	4	4	0%	67%	33%
Tai'erzhuang	Zaozhuang	5	5	6	0	4	4	4	4	5	5	3	4	9%	45%	45%
Annual Proportion of sections with Grade V or worse		50%	50%	50%	0%	20%	33%	0%	0%	17%	17%	0%	0%			

Table 1.8 Variation in water quality grades for monitoring sections of Yishusi Water System

Sections	Regions	1997	1998	1999	2001	2002	2003	2004	2005	2006	2007	2008	2009	Proportion of water quality grades for each section		
														II - III	II - III	II - III
Gonglu Bridge (310)	Xuzhou	0	5	5	0	6	6	4	4	4	4	4	4	0%	60%	40%
Aishan West Bridge	Xuzhou	0	3	6	0	6	6	5	6	6	4	3	4	20%	20%	60%
Beiwaihuan Bridge	Xuzhou	0	6	6	0	0	5	4	4	4	4	4	4	0%	67%	33%
Dongpianhong	Linyi	0	6	6	0	4	4	4	4	4	4	4	5	0%	70%	30%
Gangshang	Linyi	0	4	4	0	4	4	4	3	3	3	3	3	50%	50%	0%
Gaofengtou	Linyi	5	2	4	0	4	4	4	4	4	4	4	4	9%	82%	9%
Huangzhuang	Jining	0	6	6	0	6	6	6	6	6	6	6	6	0%	0%	100%
Jiaoyi	Linyi	0	6	5	0	0	4	4	4	3	3	3	4	33%	44%	22%
Jiezhuang	Linyi	0	5	6	0	0	4	4	4	4	4	4	4	0%	78%	22%
Liji	Xuzhou	4	3	5	6	5	6	5	4	3	3	3	3	42%	17%	42%
Daxing Bridge	Linyi	0	6	6	0	4	5	4	4	4	4	4	4	0%	70%	30%
Malou	Jining	0	6	6	0	0	6	5	5	6	6	4	4	0%	22%	78%
Qunle Bridge	Jining	0	6	6	0	0	6	6	6	6	6	5	4	0%	11%	89%
Shagou Bridge	Linyi	0	6	5	0	0	5	4	4	4	4	4	4	0%	67%	33%
Shuyuan	Jining	0	6	6	0	0	6	6	6	6	5	4	5	0%	11%	89%
Xiyao	Jining	5	5	6	0	0	4	5	5	4	4	4	4	0%	50%	50%
Yulou	Heze	0	6	6	0	0	6	6	6	6	6	6	5	0%	0%	100%
Zhangtuan Bridge	Linyi	0	3	6	0	6	6	6	4	4	4	4	4	10%	50%	40%
Zhangzhuang	Suqian	4	3	2	3	2	3	3	3	3	3	3	3	92%	8%	0%
Annual Proportionof sections with Grade Vor worse		50%	68%	84%	50%	50%	63%	47%	37%	32%	26%	16%	21%			

References

Chinese Center for Disease Control and Prevention. Program on cancer control and prevention in Huai River Basin, vol. I, The report on retrospective survey on death causes. Beijing: Peking Union Medical College Press; 2007. p. 4. ISBN 978-7-81136-163-6

The Editorial Committee. Atlas of Cancer Mortality in the People's Republic of China. Beijing: China Map Press; 1979.

Gu H, Shen H, Wu G. Characteristics of water resource allocation and implementation measures in Huaihe River Basin. 2006. http://www.hrc.gov.cn/detail?documentid=22475

MOH, P. R. China, the Report on the third national retrospective survey on the death causes by sampling, edited by Chen Zhu. Beijing: Peking Union Medical College Press; 2008. ISBN 9787811360769.

Song D, Chen W, Gao Y. Reasonability of nitrogen and pesticide use in Huaihe River Basin and its environmental impact [J]. J Agri Environ Sci. 2011;30(6):1144–51.

Tao Z. Study on burden of water-related disease in "cancer village" in Huaihe River Basin—discussion of the burden of disease attributed to environmental pollutants. Beijing: Chinese Center for Disease Control and Prevention; 2010.

Zhang Z, Li L, chief editor. Atlas of China's groundwater resources and environment. Beijing: China Map Publishing House; 2004.

Zhu W. A decade of Huai River wastewater management. Oriental Outlook Weekly, 2004–11–02. http://news.jschina.com.cn

2 Parent Maps

G. Yang, D. Zhuang (eds.), *Atlas of the Huai River Basin Water Environment: Digestive Cancer Mortality*,
DOI 10.1007/978-94-017-8619-5_2,

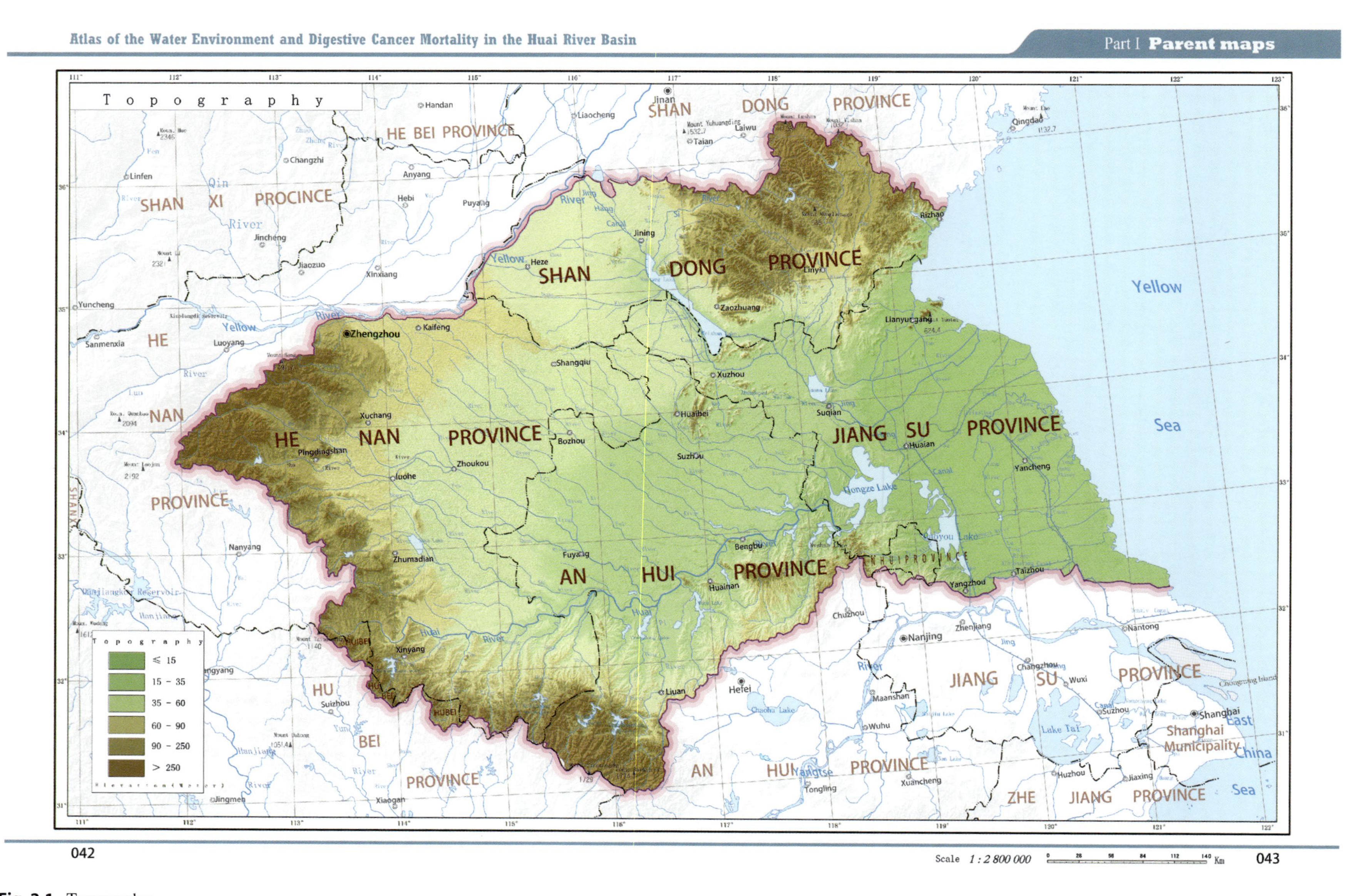

Fig. 2.1 Topography

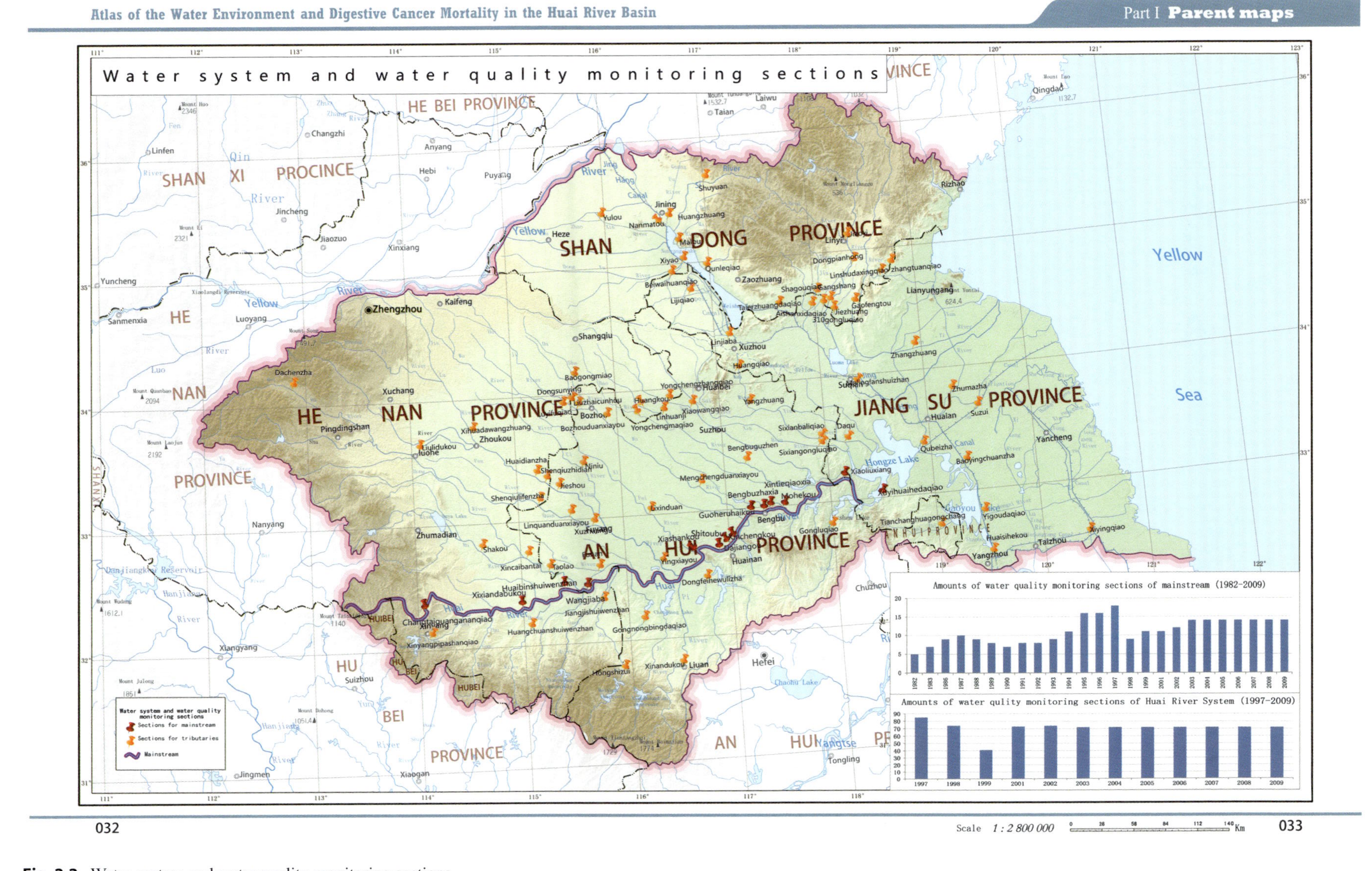

032

Scale 1 : 2 800 000 Km

033

Fig. 2.2 Water system and water quality monitoring sections

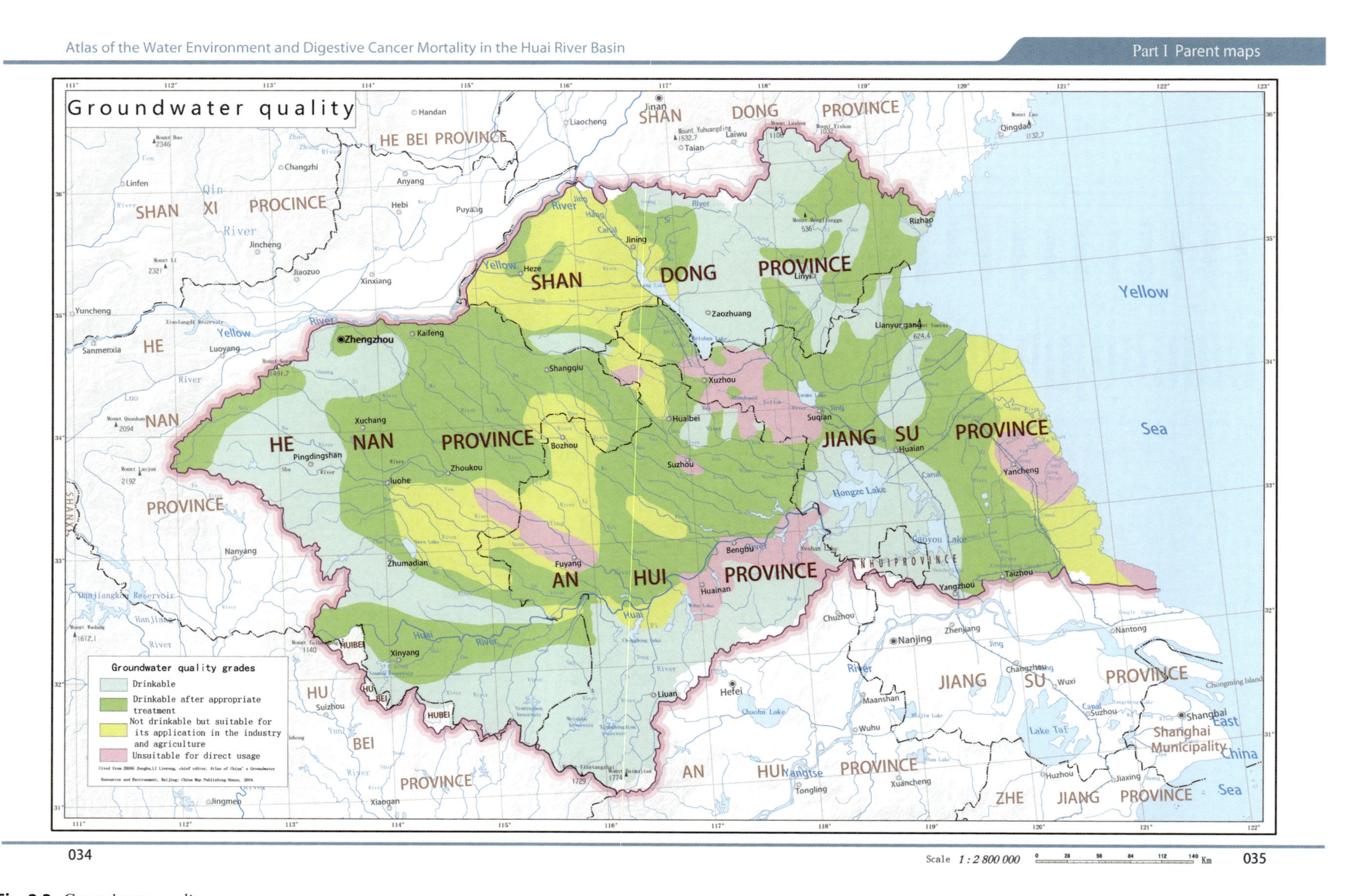

Fig. 2.3 Groundwater quality

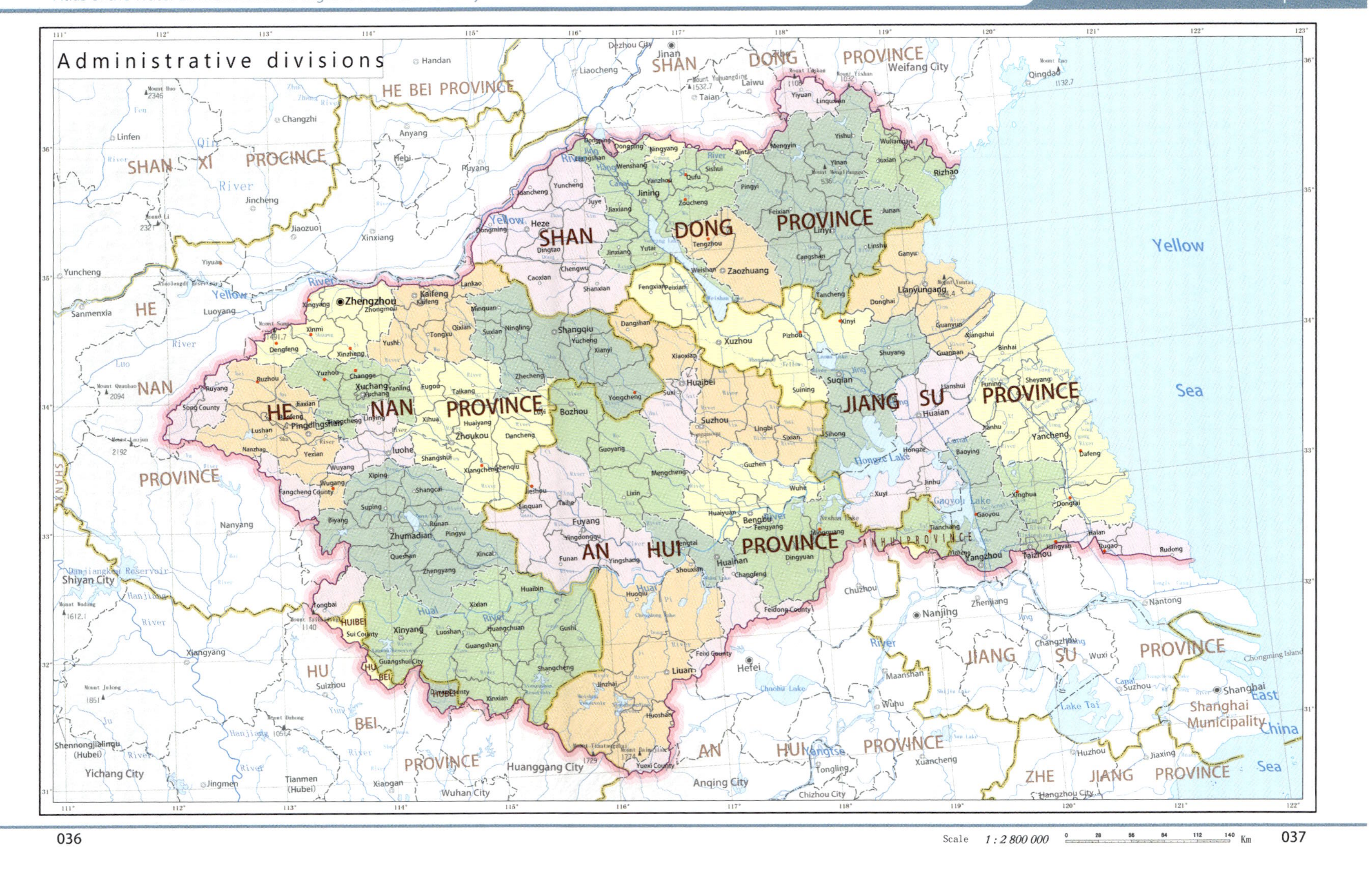

Fig. 2.4 Administrative divisions

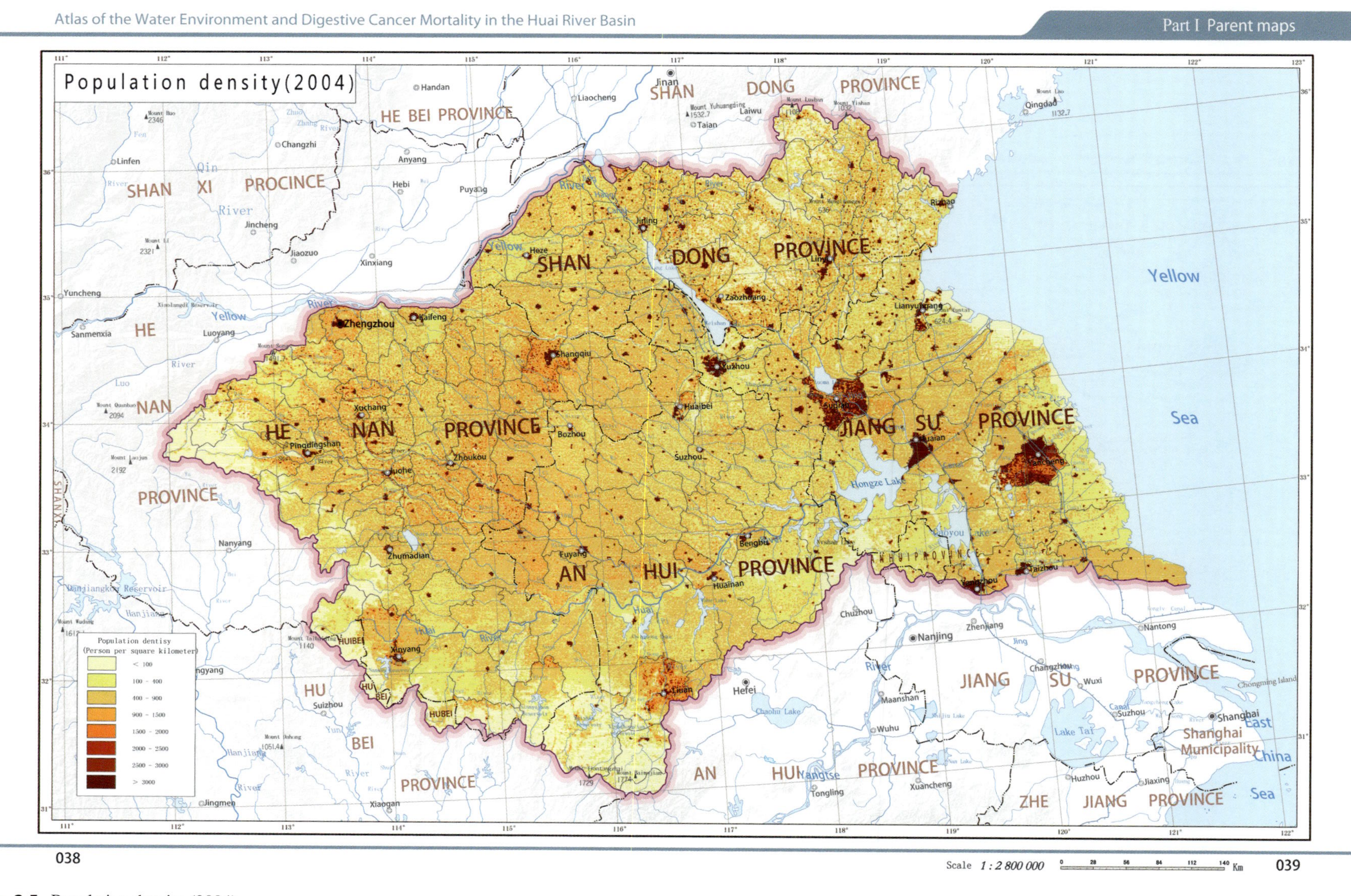

Fig. 2.5 Population density (2004)

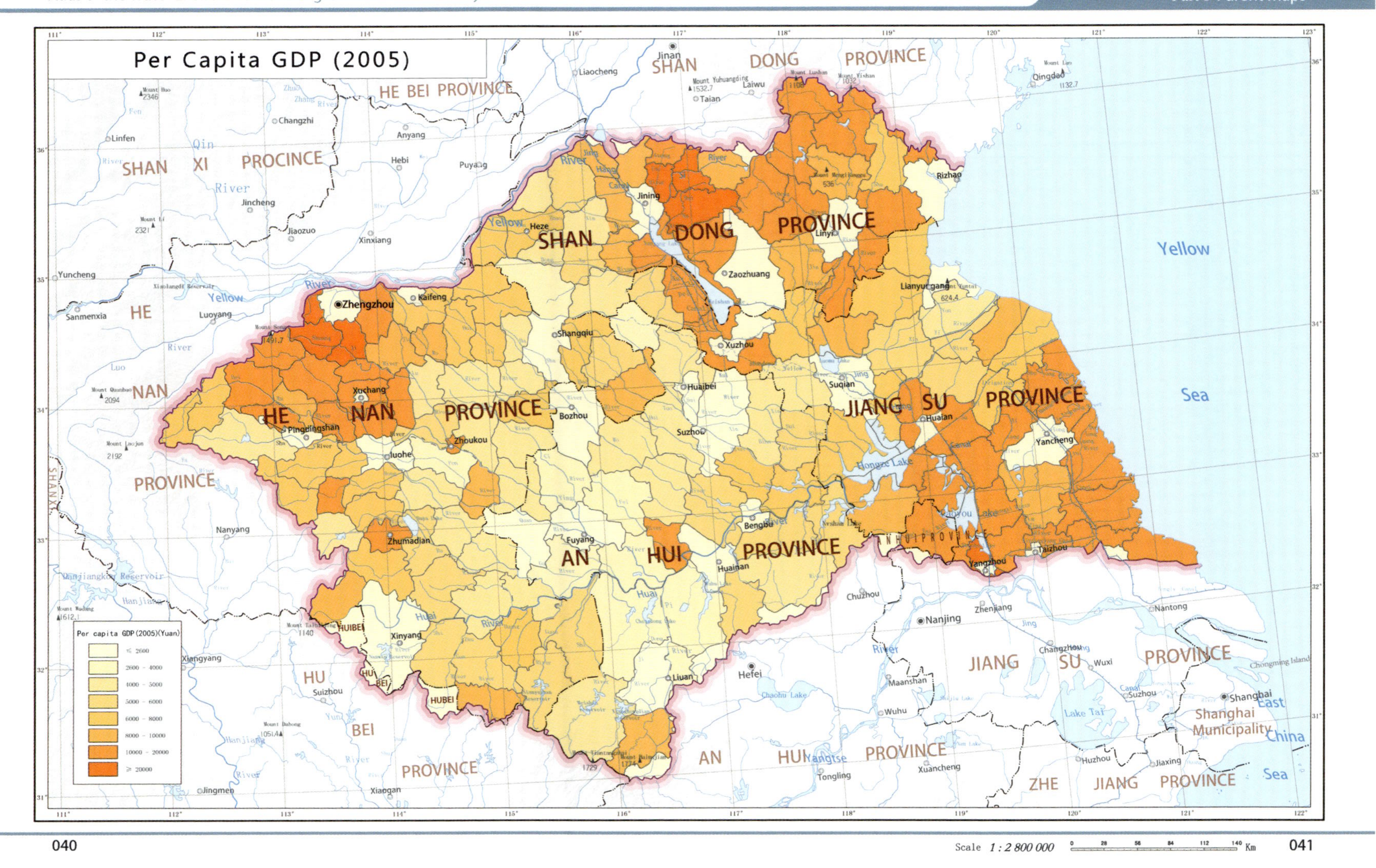

Fig. 2.6 Per Capita GDP (2005)

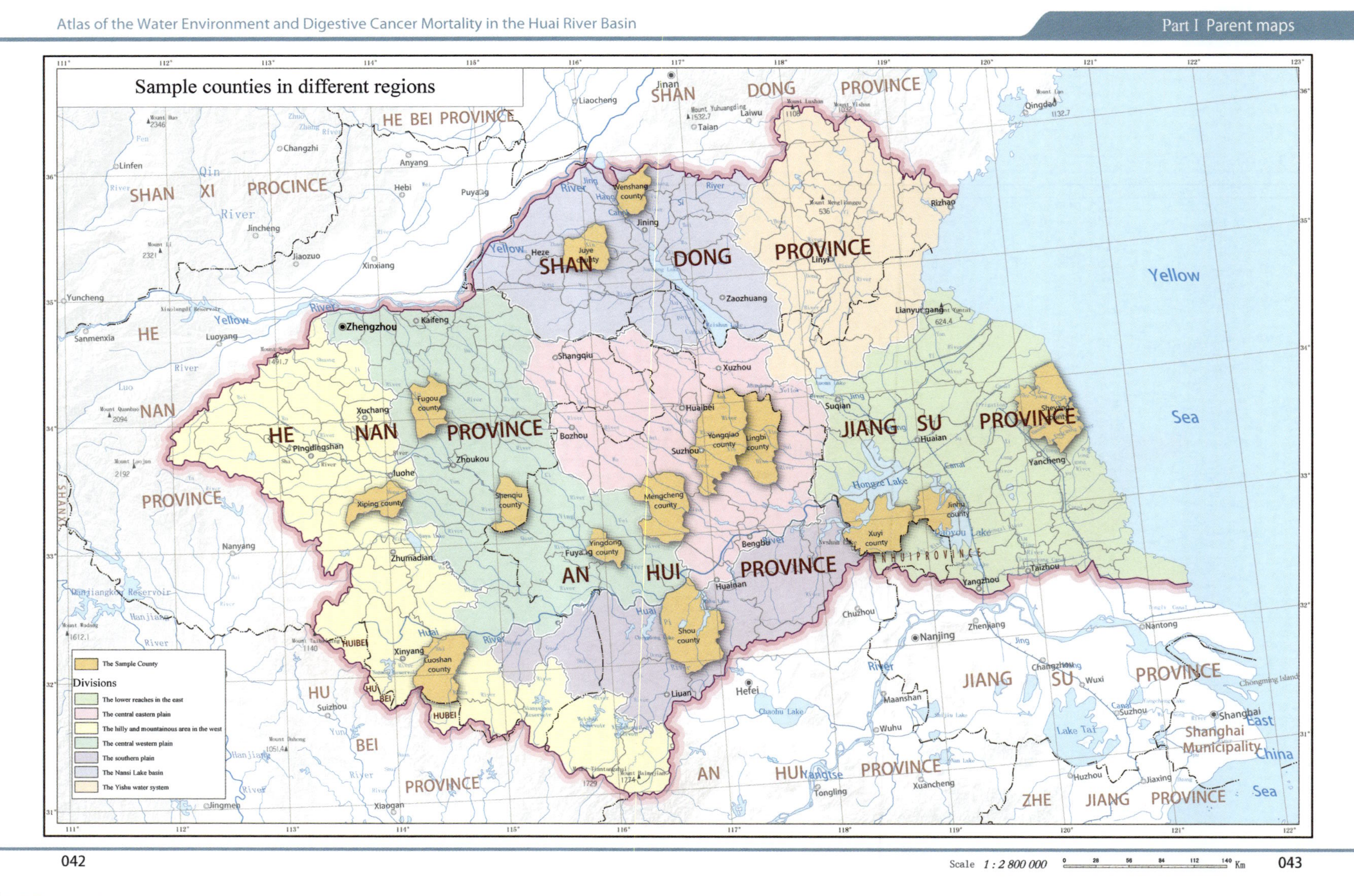

Fig. 2.7 Sample counties in different regions

3 Variation in the Main Stream Water Environment

G. Yang, D. Zhuang (eds.), *Atlas of the Huai River Basin Water Environment: Digestive Cancer Mortality*,
DOI 10.1007/978-94-017-8619-5_3,

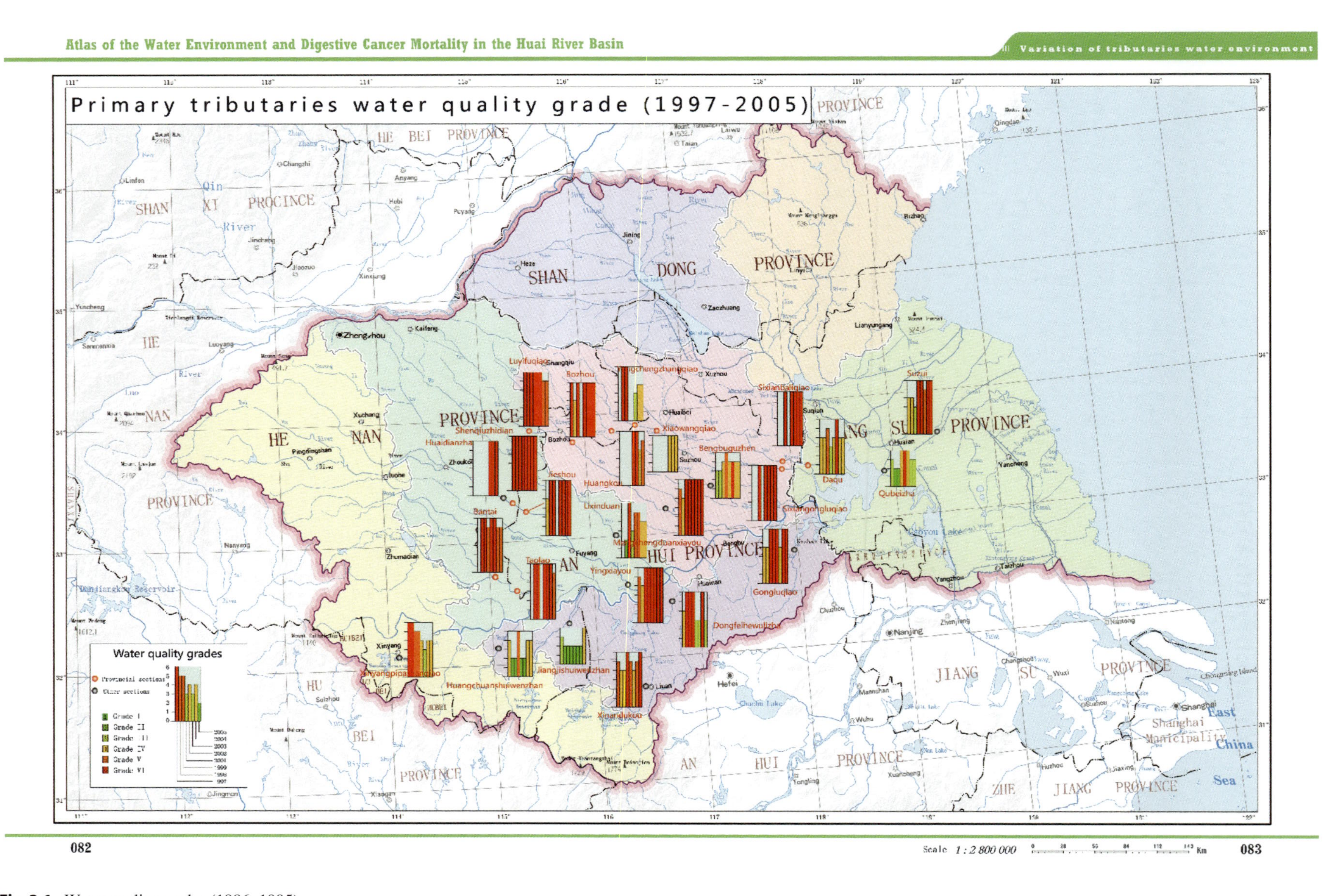

Fig. 3.1 Water quality grades (1986–1995)

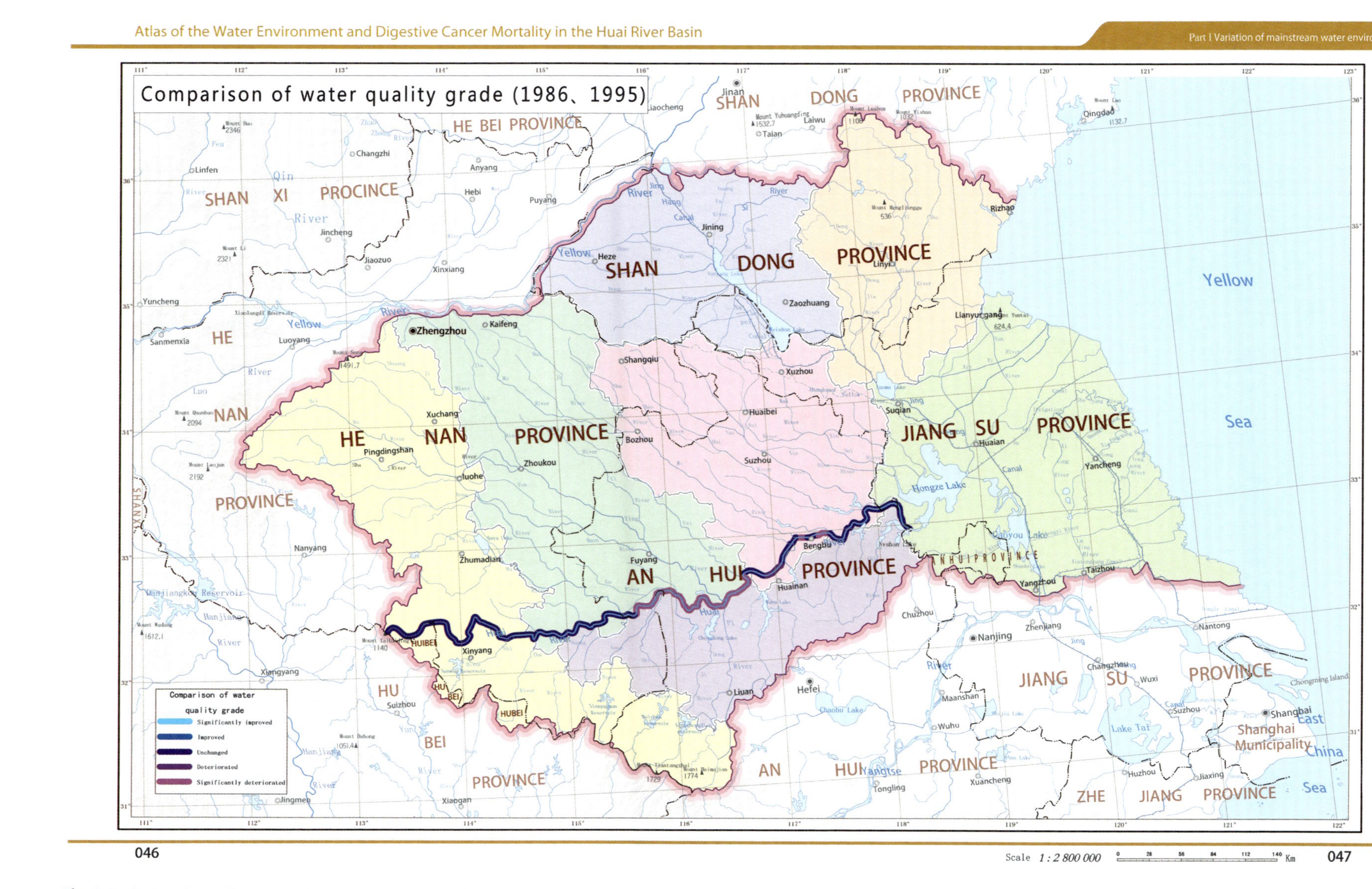

Fig. 3.2 Comparison of water quality grades (1986–1995)

Fig. 3.3 BOD concentration grades (1986–1995)

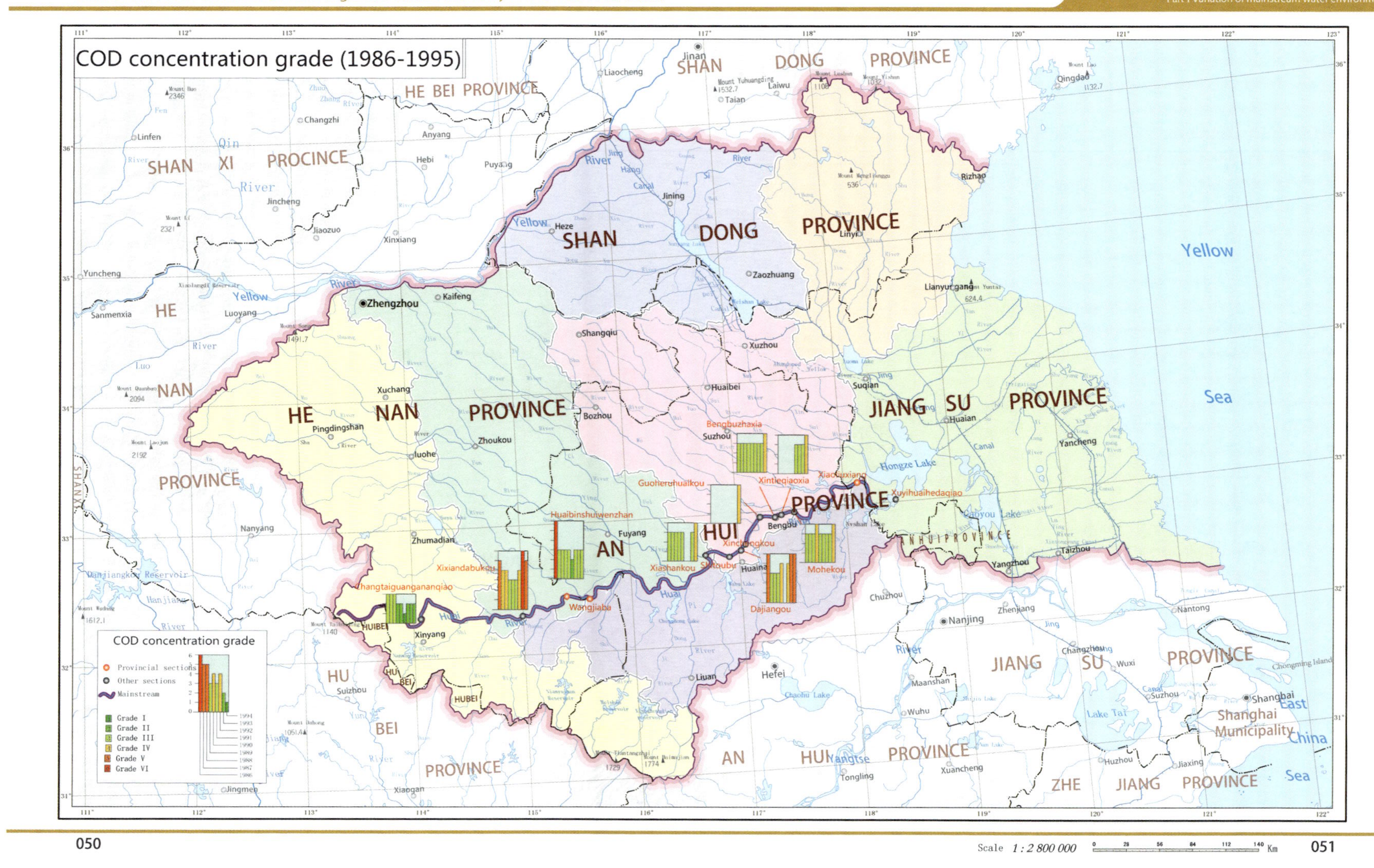

Fig. 3.4 COD concentration grades (1986–1995)

Fig. 3.5 Ammonia nitrogen concentration grades (1986–1995)

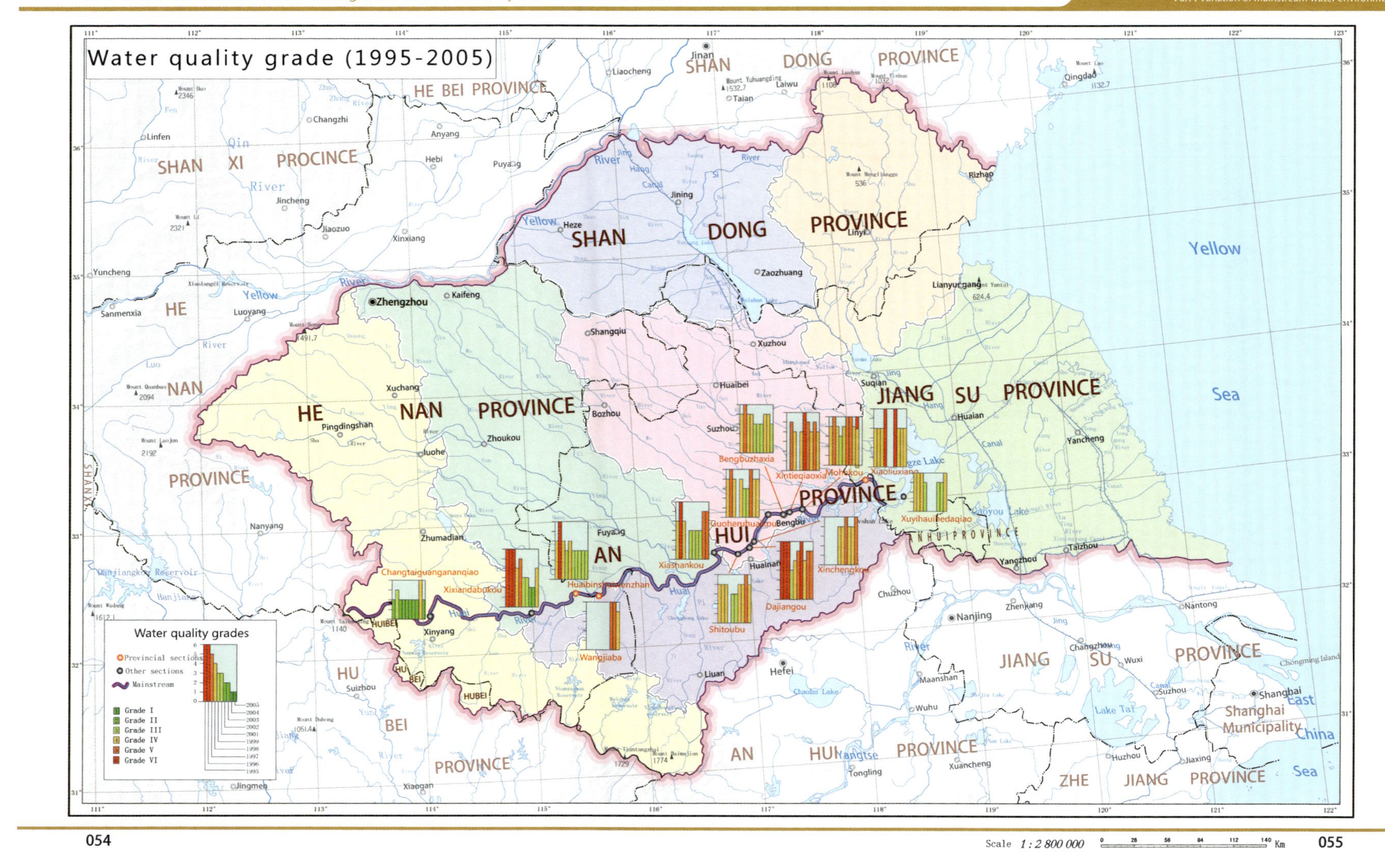

Fig. 3.6 Water quality grades (1995–2005)

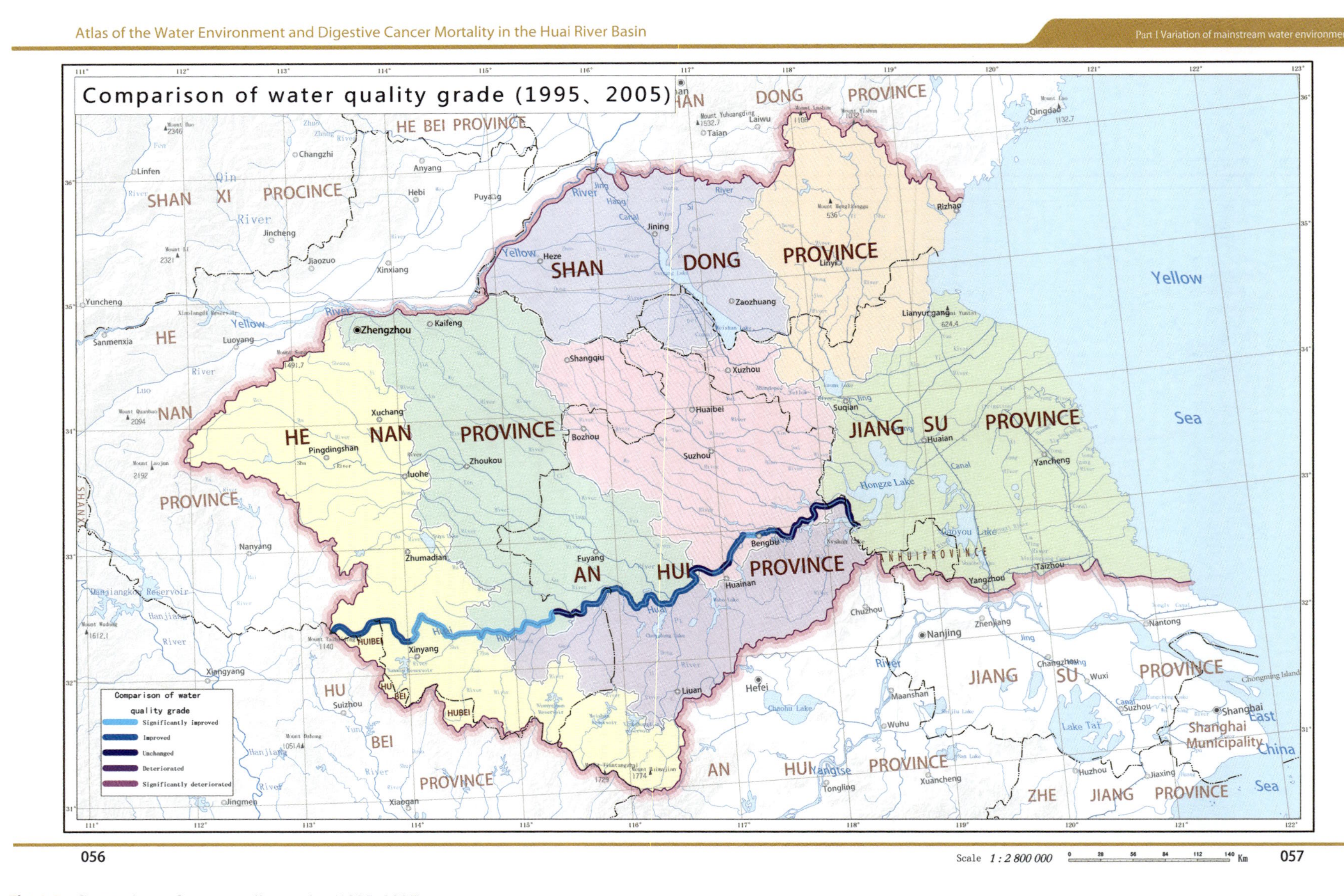

Fig. 3.7 Comparison of water quality grades (1995–2005)

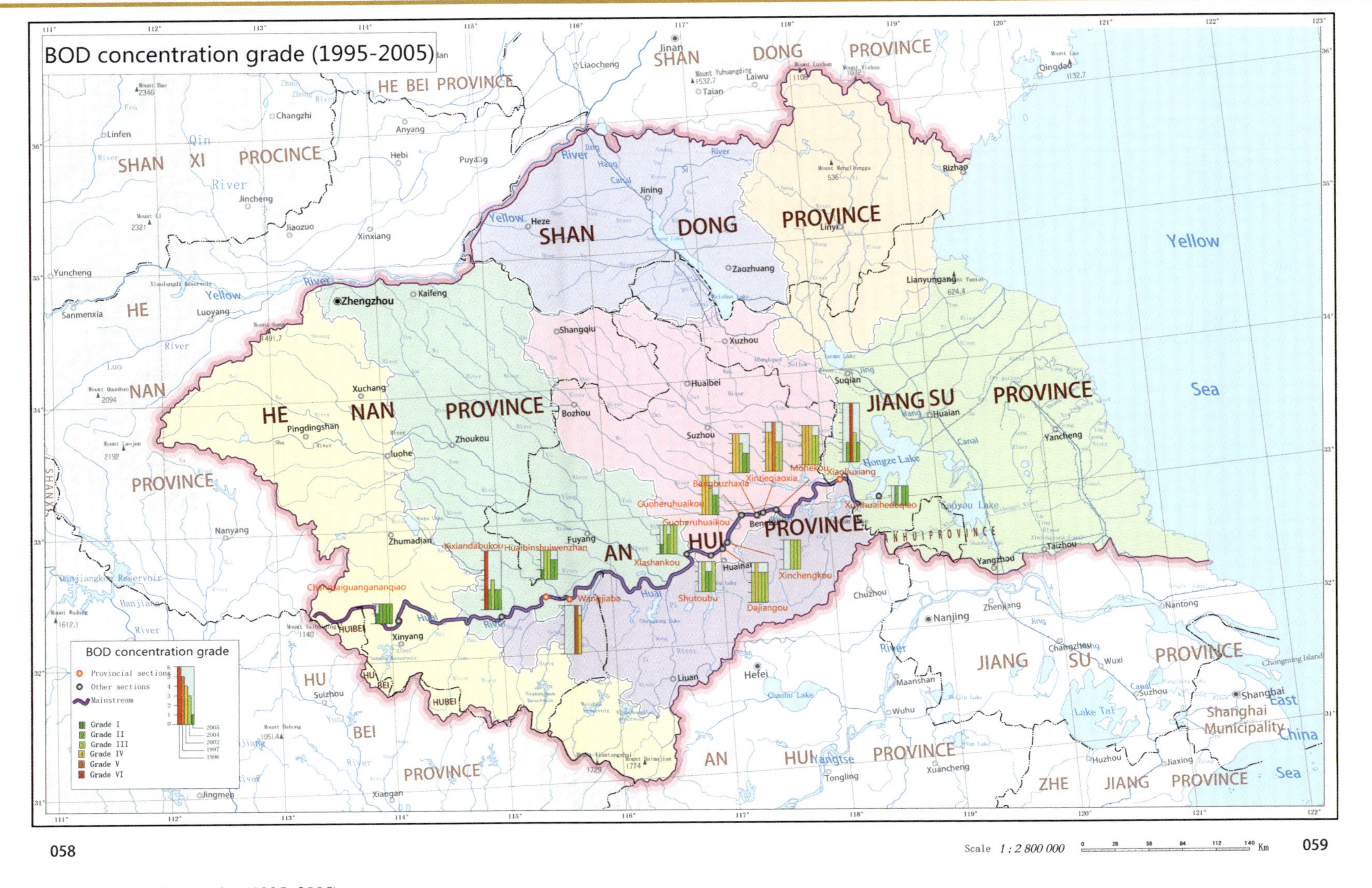

Fig. 3.8 BOD concentration grades (1995–2005)

Fig. 3.9 COD concentration grades (1995–2005)

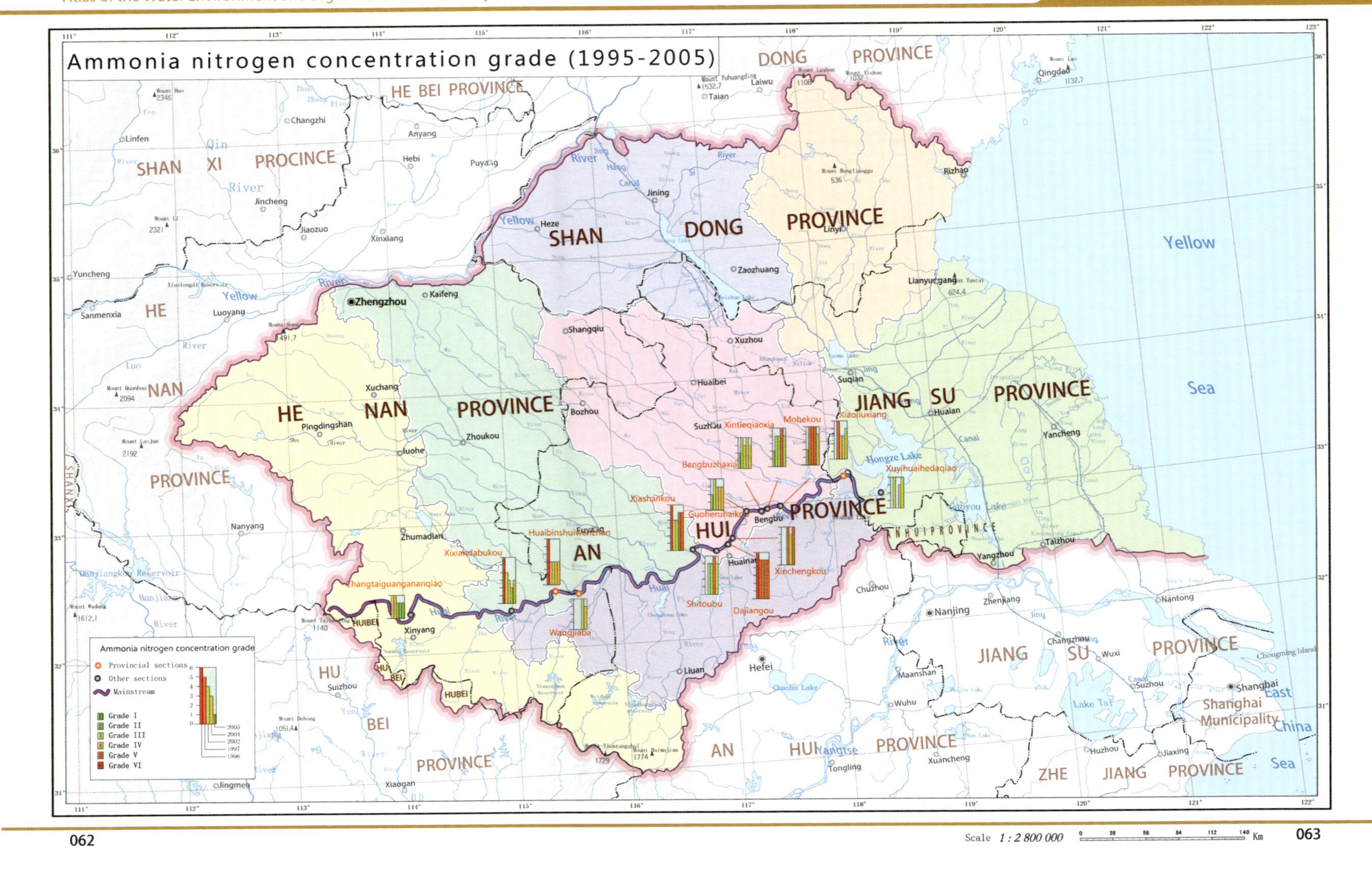

Fig. 3.10 Ammonia nitrogen concentration grades (1995–2005)

Fig. 3.11 Water quality grades (2005–2009)

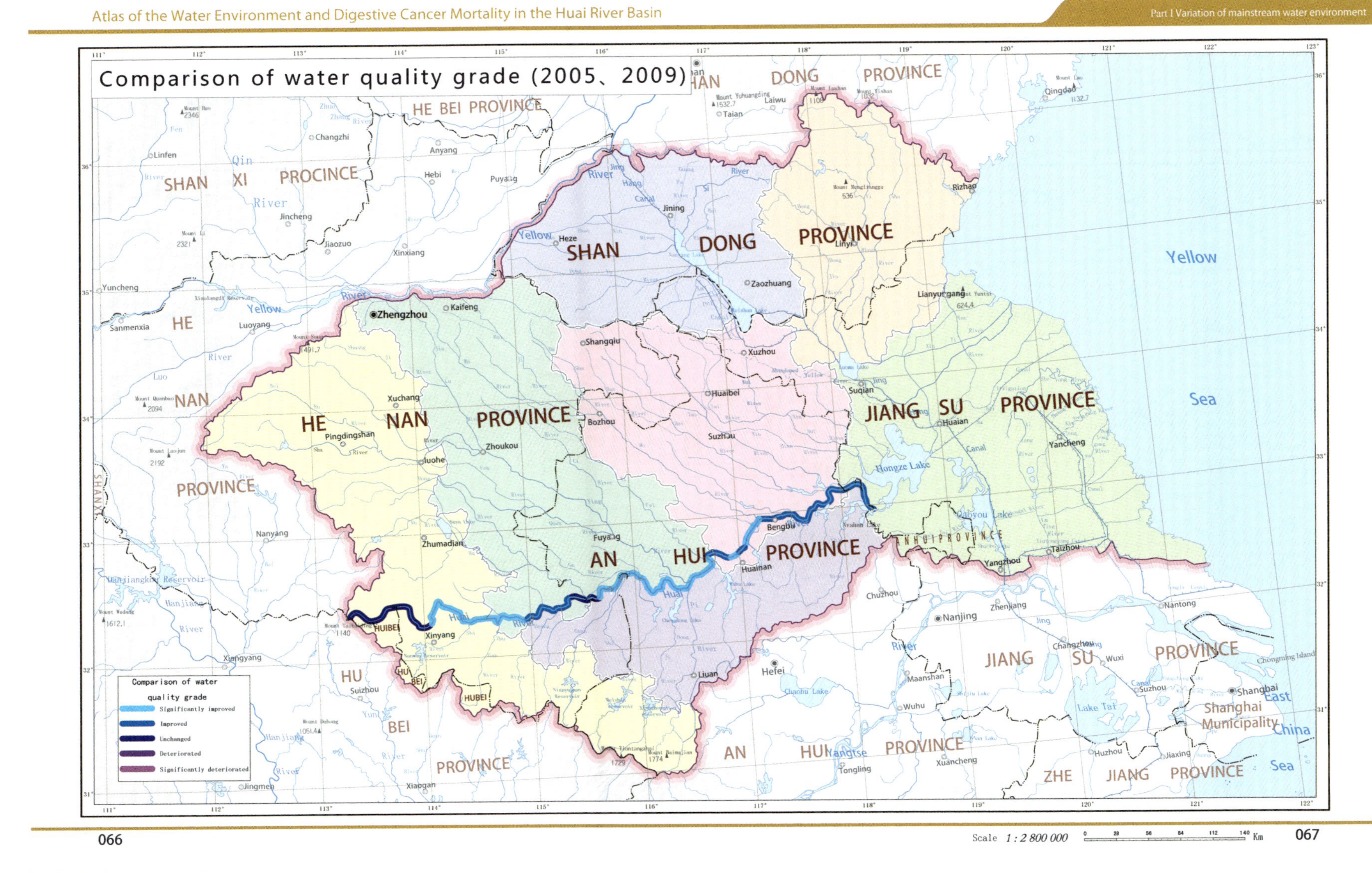

Fig. 3.12 Comparison of water quality grades (2005–2009)

Fig. 3.13 BOD concentration grades (2005–2009)

Fig. 3.14 COD concentration grades (2005–2009)

Fig. 3.15 Ammonia nitrogen concentration grades (2005–2009)

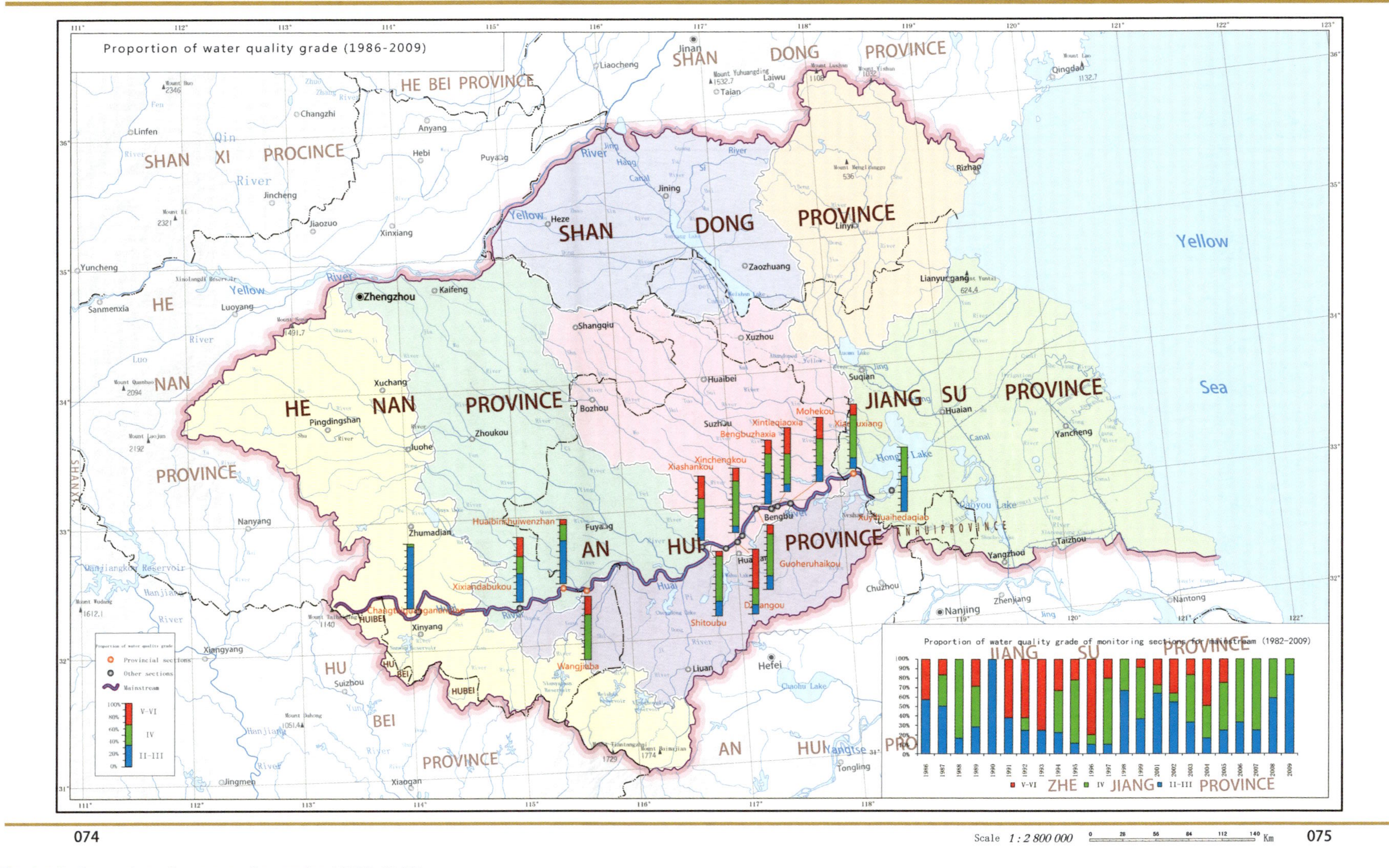

Fig. 3.16 Proportion of water quality grades (1986–2009)

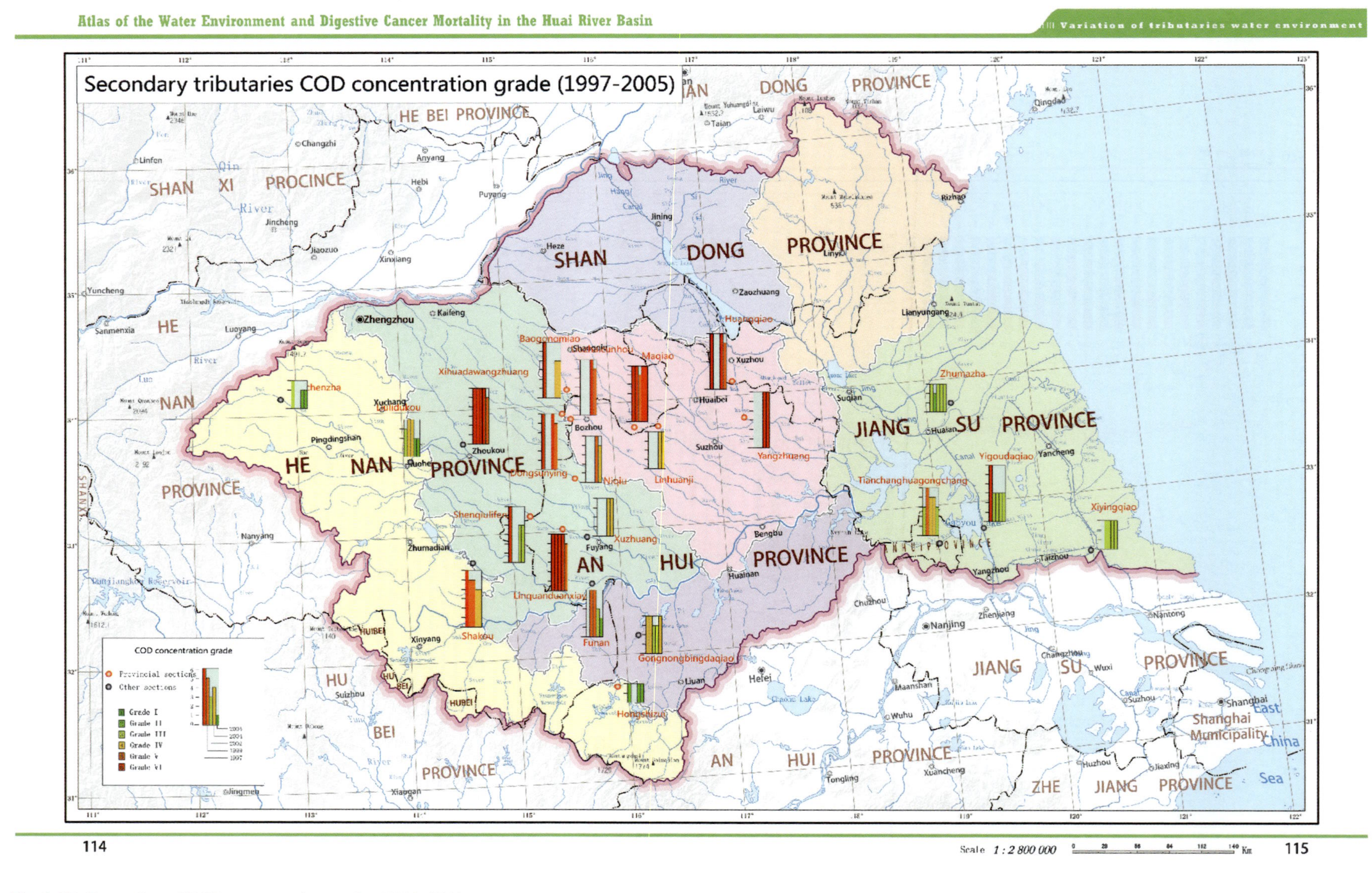

Fig. 3.17 Proportion of BOD concentration grades (1982–2009)

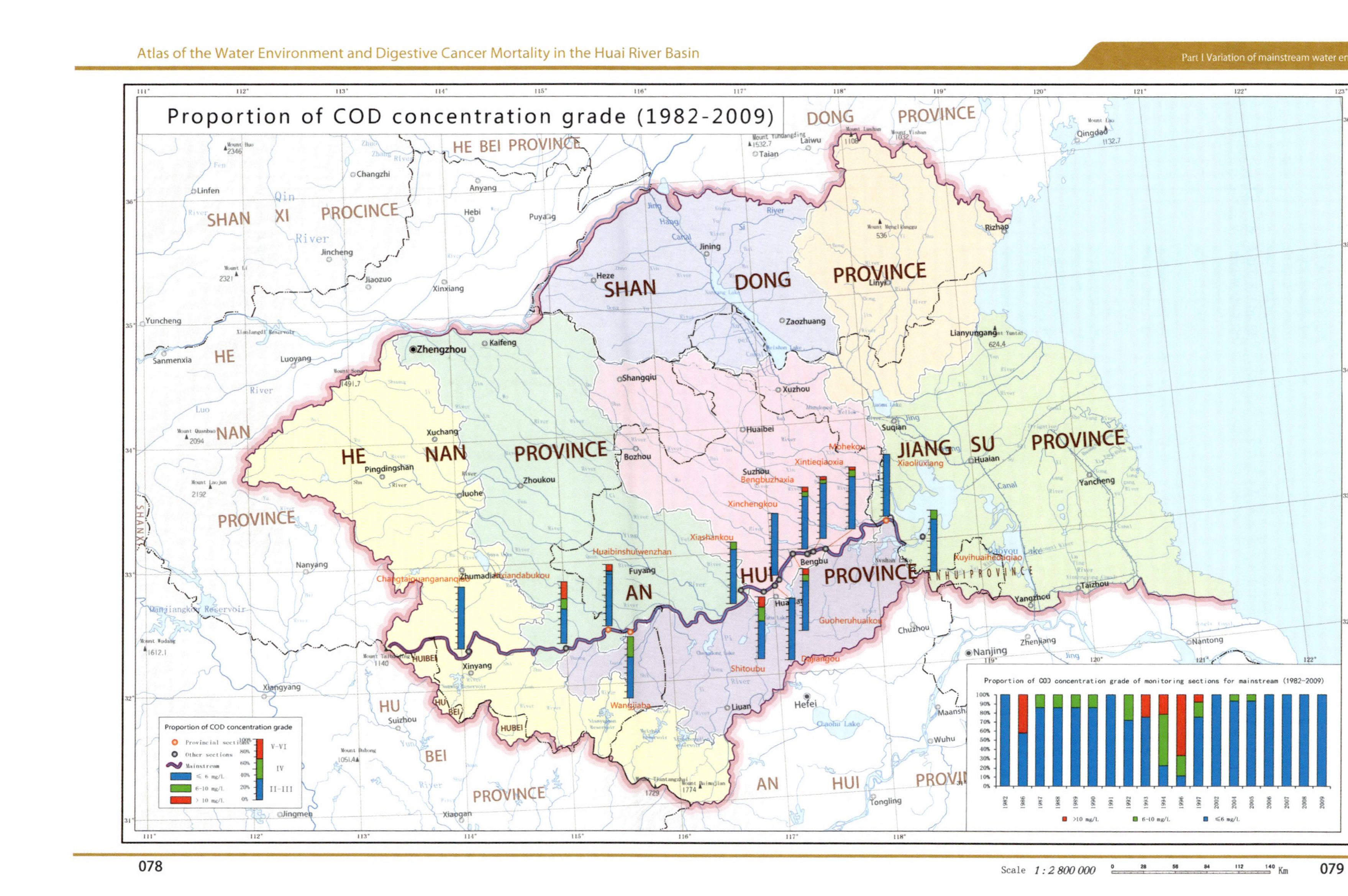

Fig. 3.18 Proportion of COD concentration grades (1982–2009)

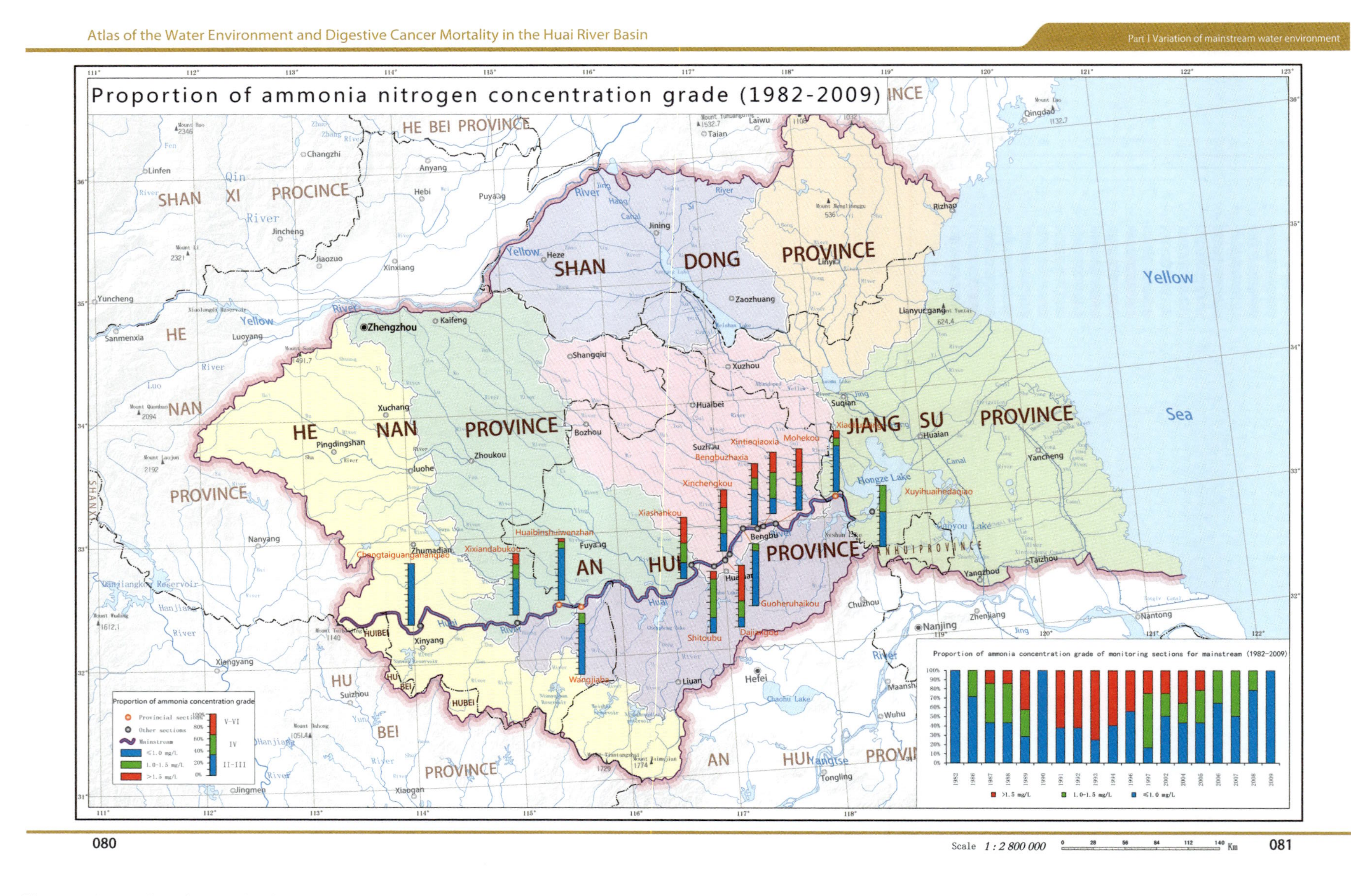

Fig. 3.19 Proportion of ammonia nitrogen concentration grades (1982–2009)

4 Variation in the Water Environment of the Tributaries

G. Yang, D. Zhuang (eds.), *Atlas of the Huai River Basin Water Environment: Digestive Cancer Mortality*,
DOI 10.1007/978-94-017-8619-5_4,

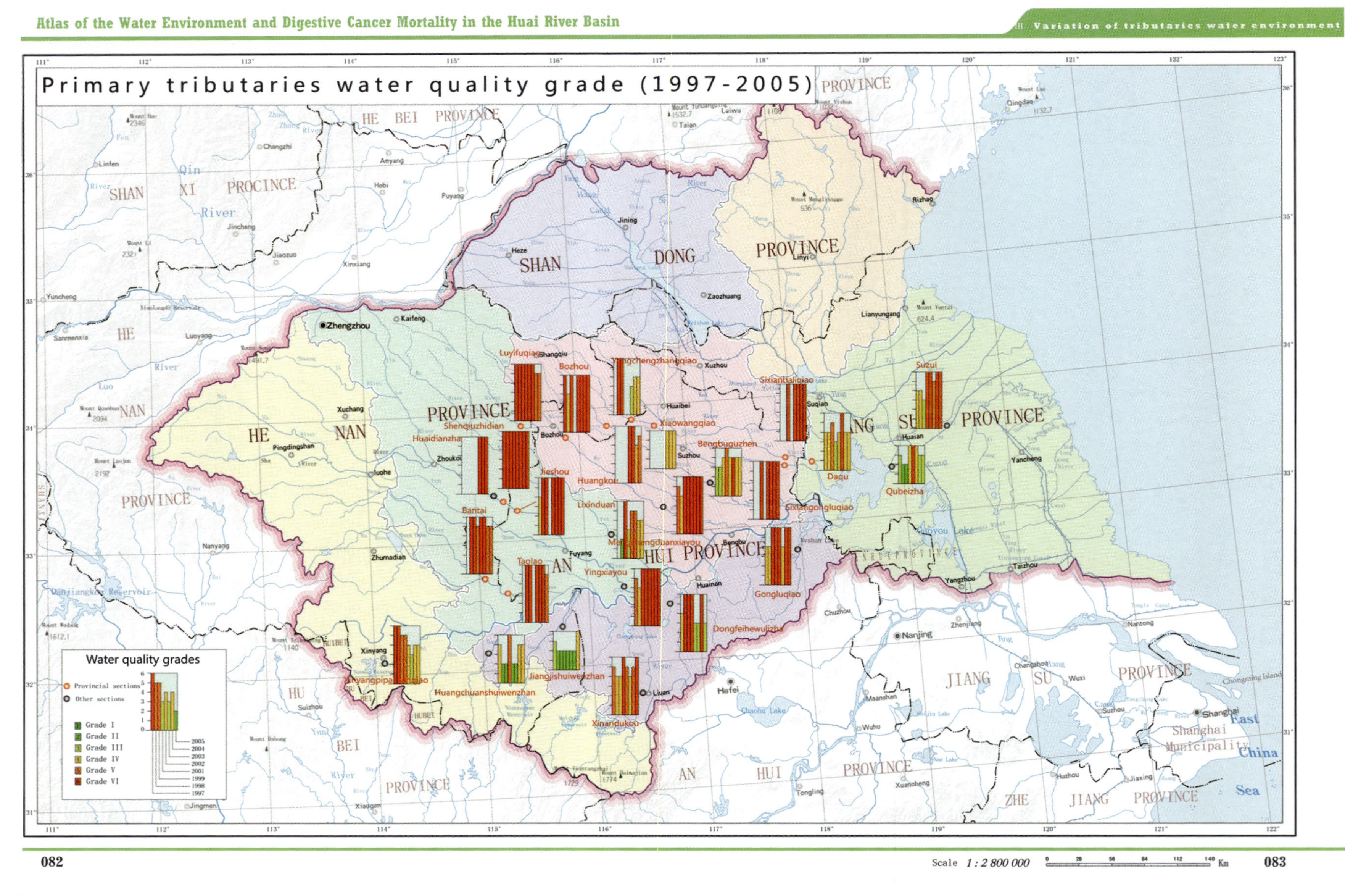

Fig. 4.1 Primary tributaries water quality grades (1997–2005)

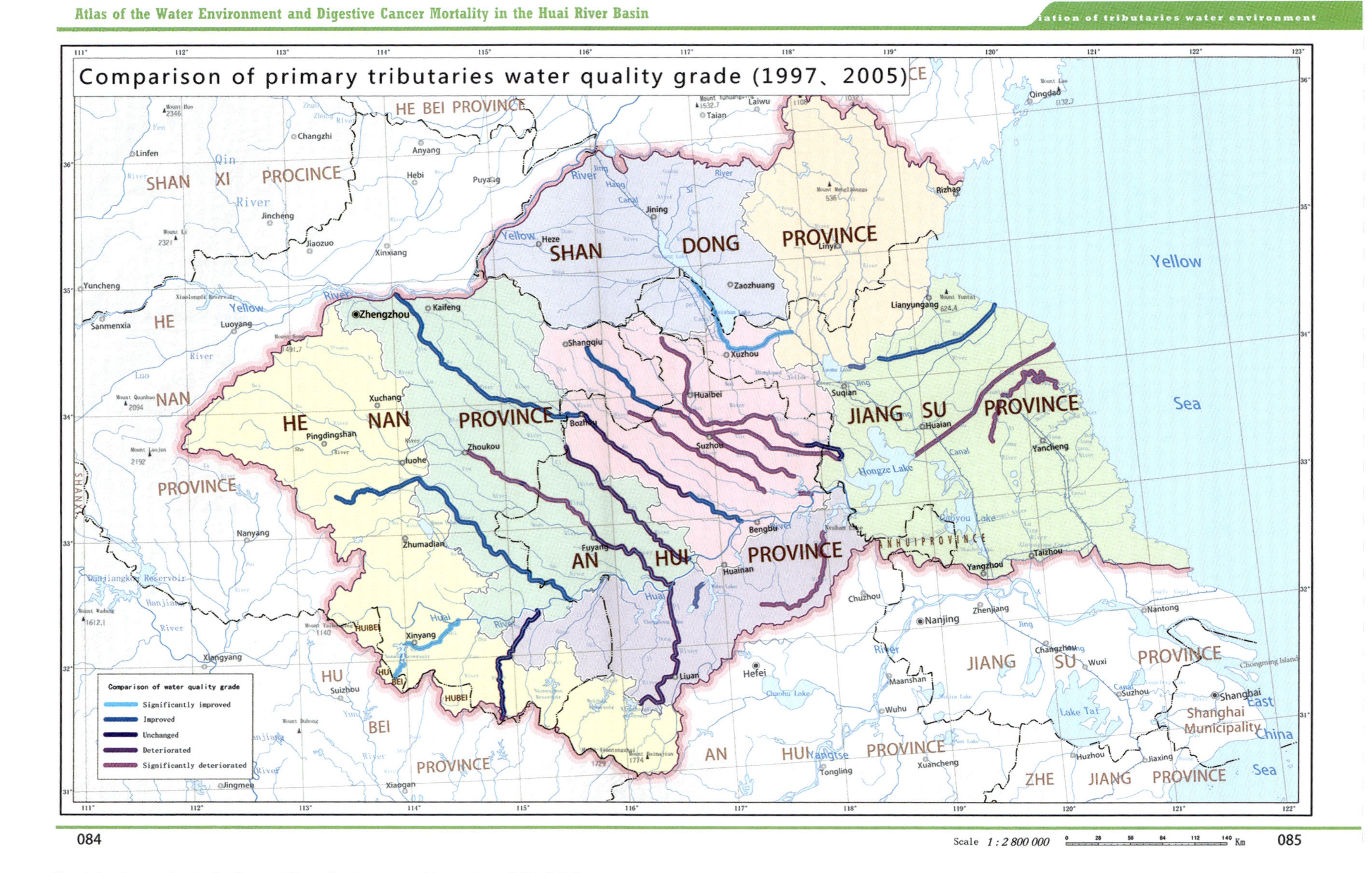

Fig. 4.2 Comparison of primary tributaries water quality grades (1997–2005)

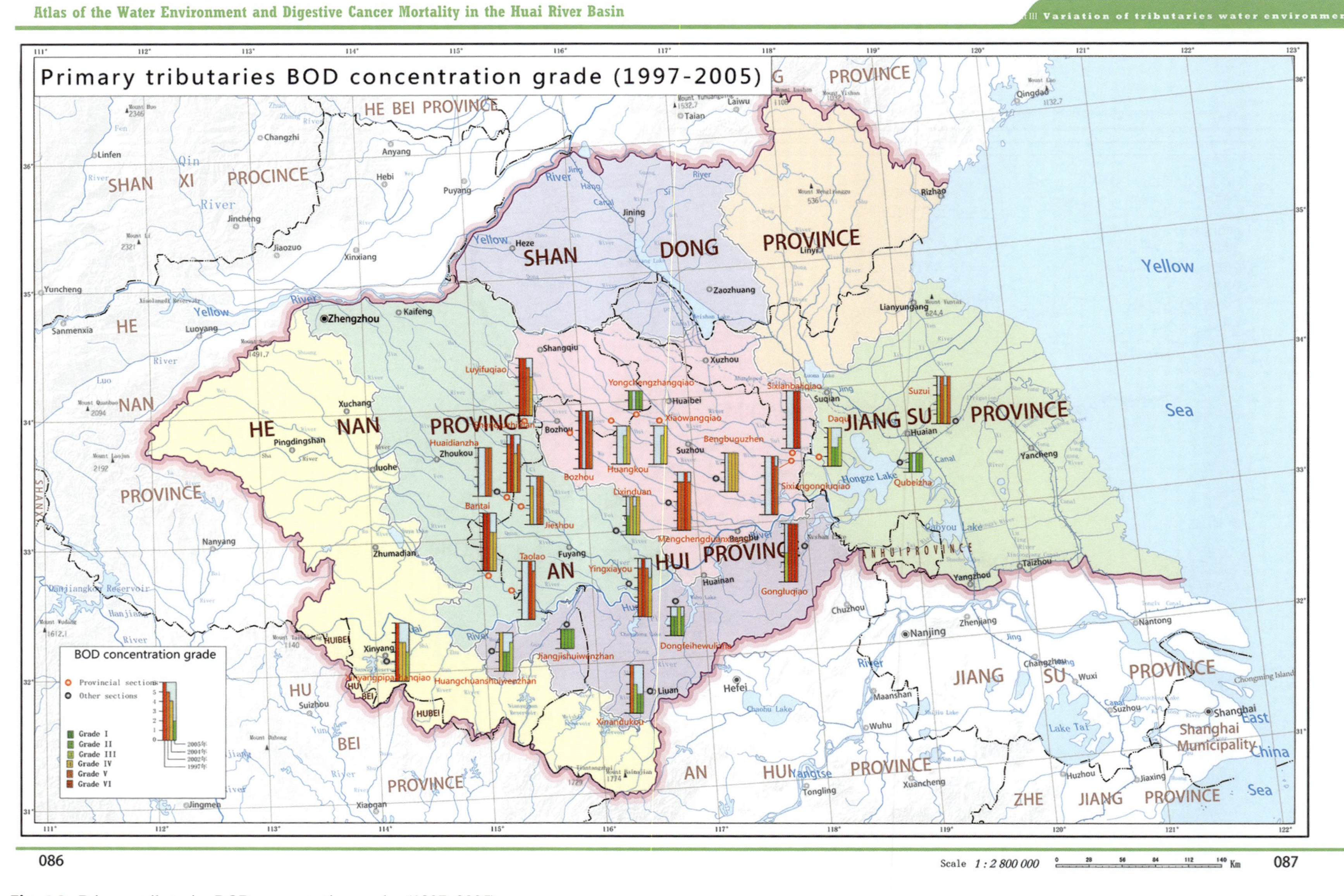

Fig. 4.3 Primary tributaries BOD concentration grades (1997–2005)

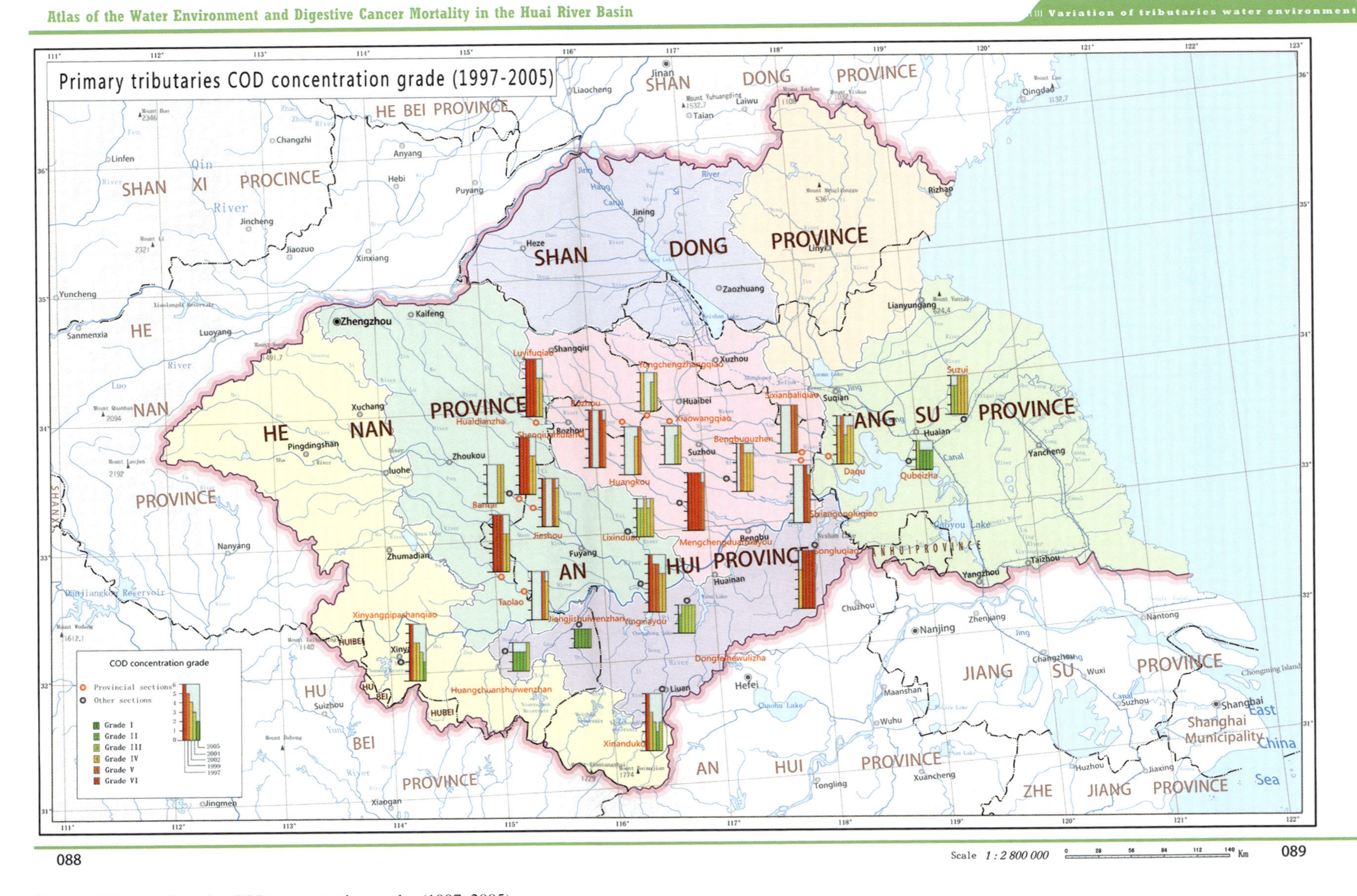

Fig. 4.4 Primary tributaries COD concentration grades (1997–2005)

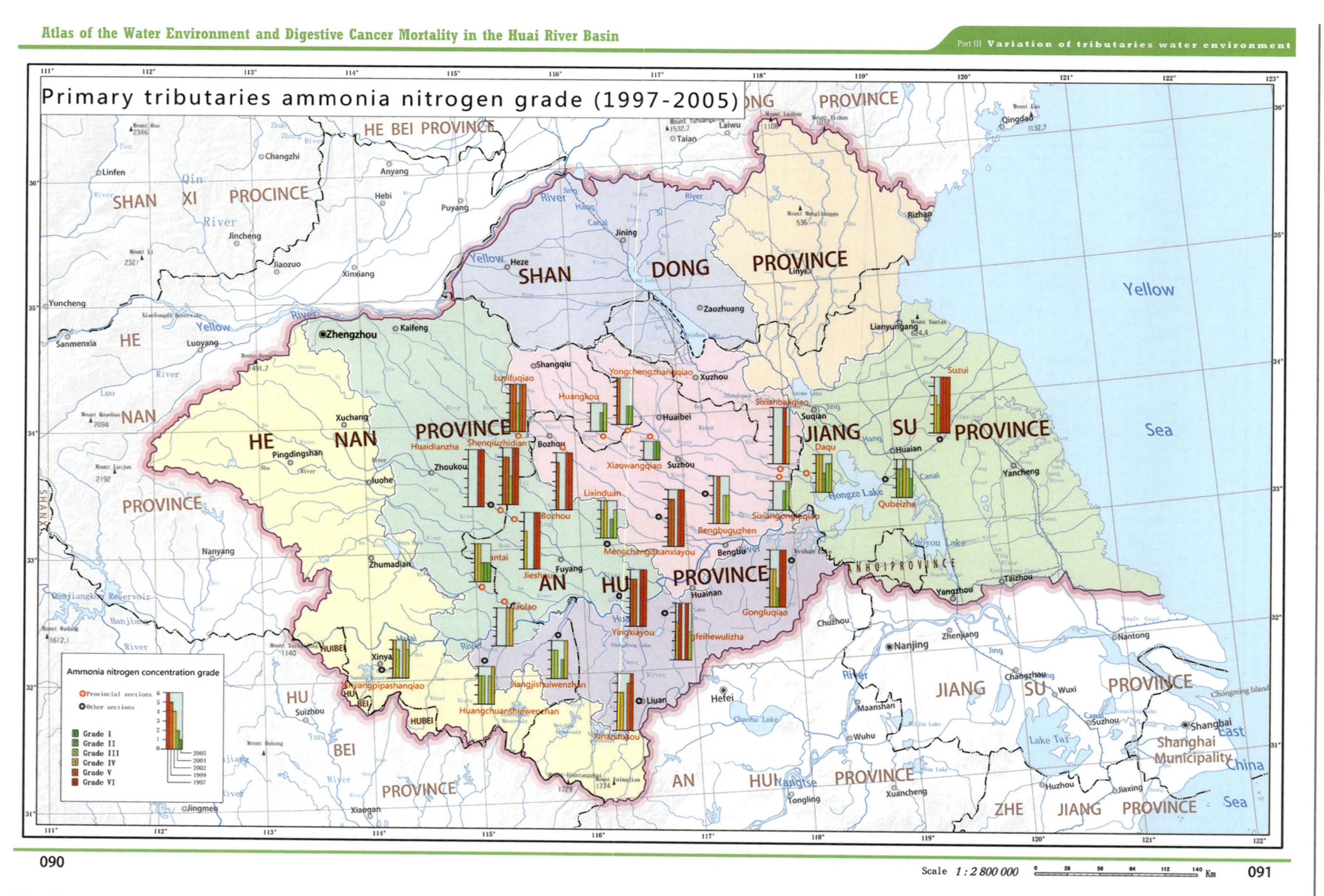

Fig. 4.5 Primary tributaries ammonia nitrogen grades (1997–2005)

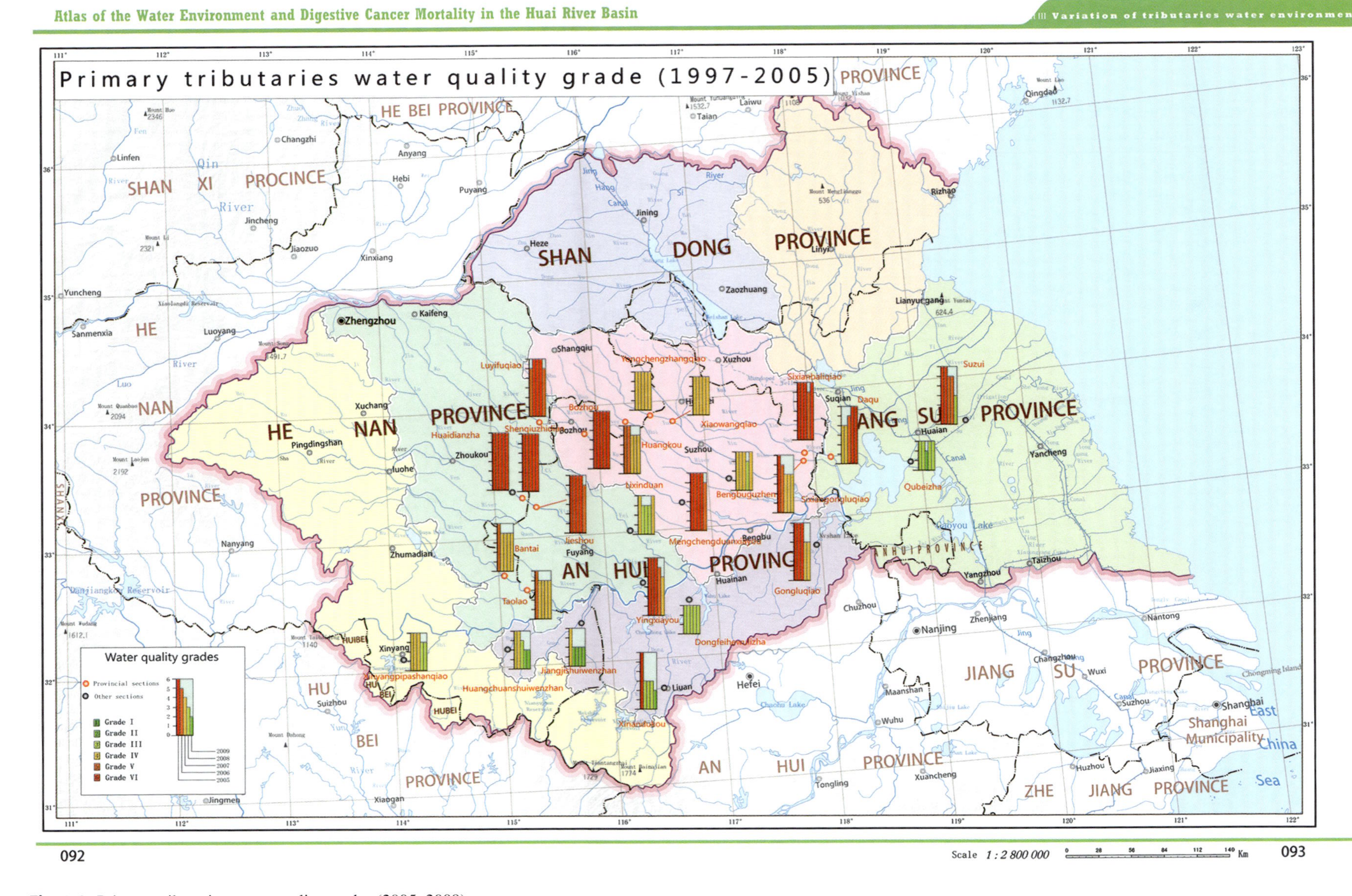

Fig. 4.6 Primary tributaries water quality grades (2005–2009)

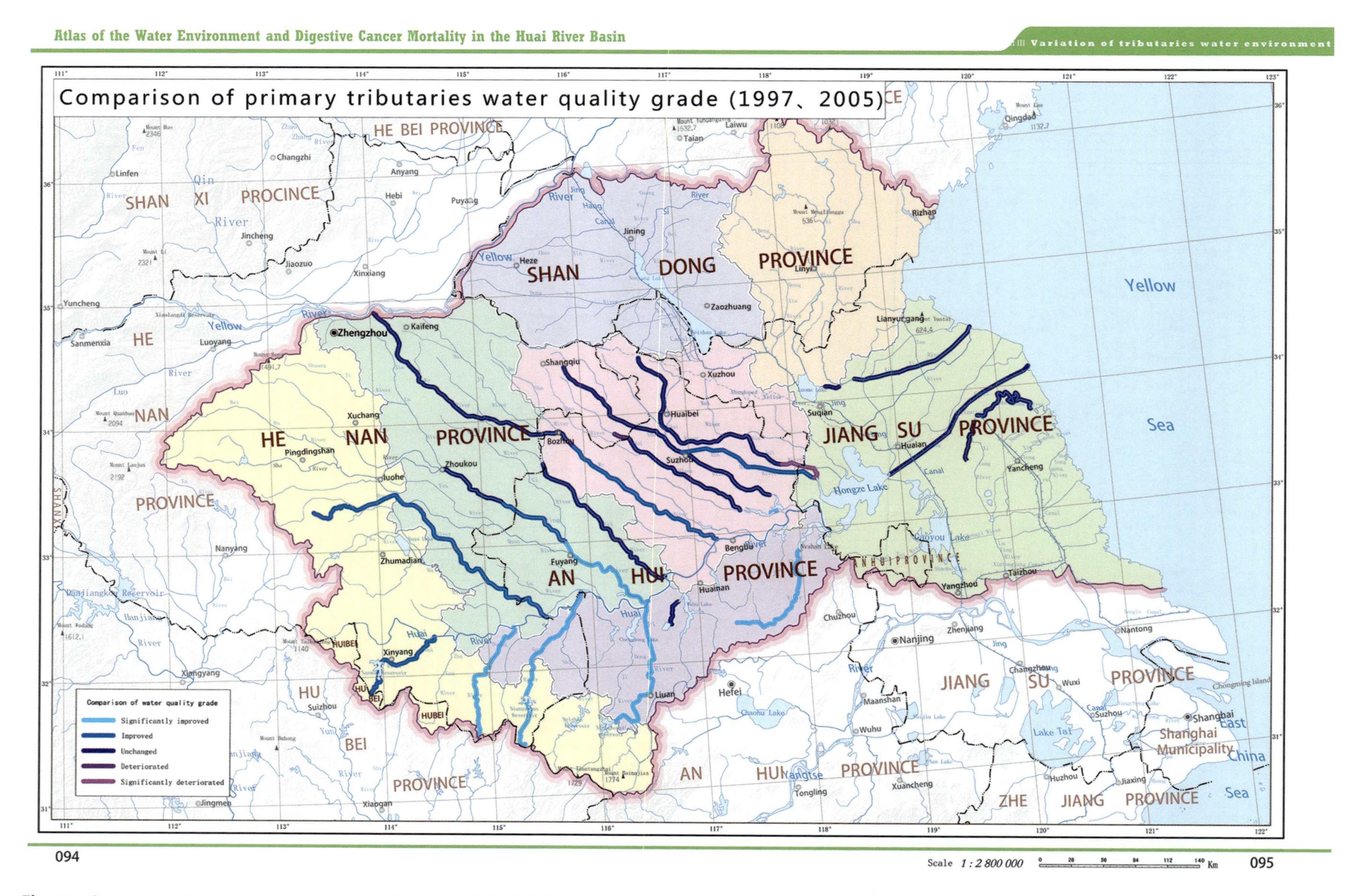

Fig. 4.7 Comparison of primary tributaries water quality grades (2005–2009)

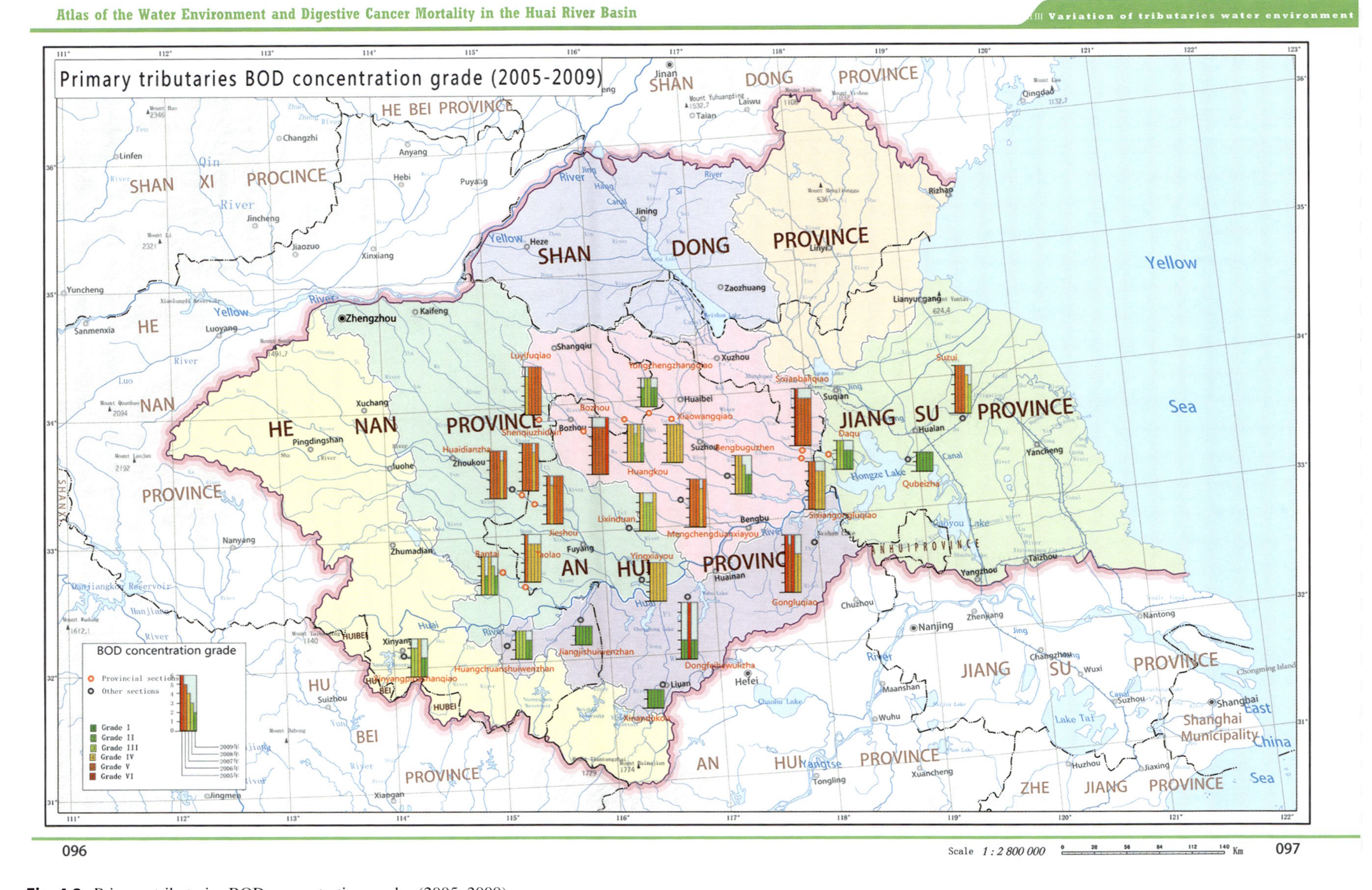

Fig. 4.8 Primary tributaries BOD concentration grades (2005–2009)

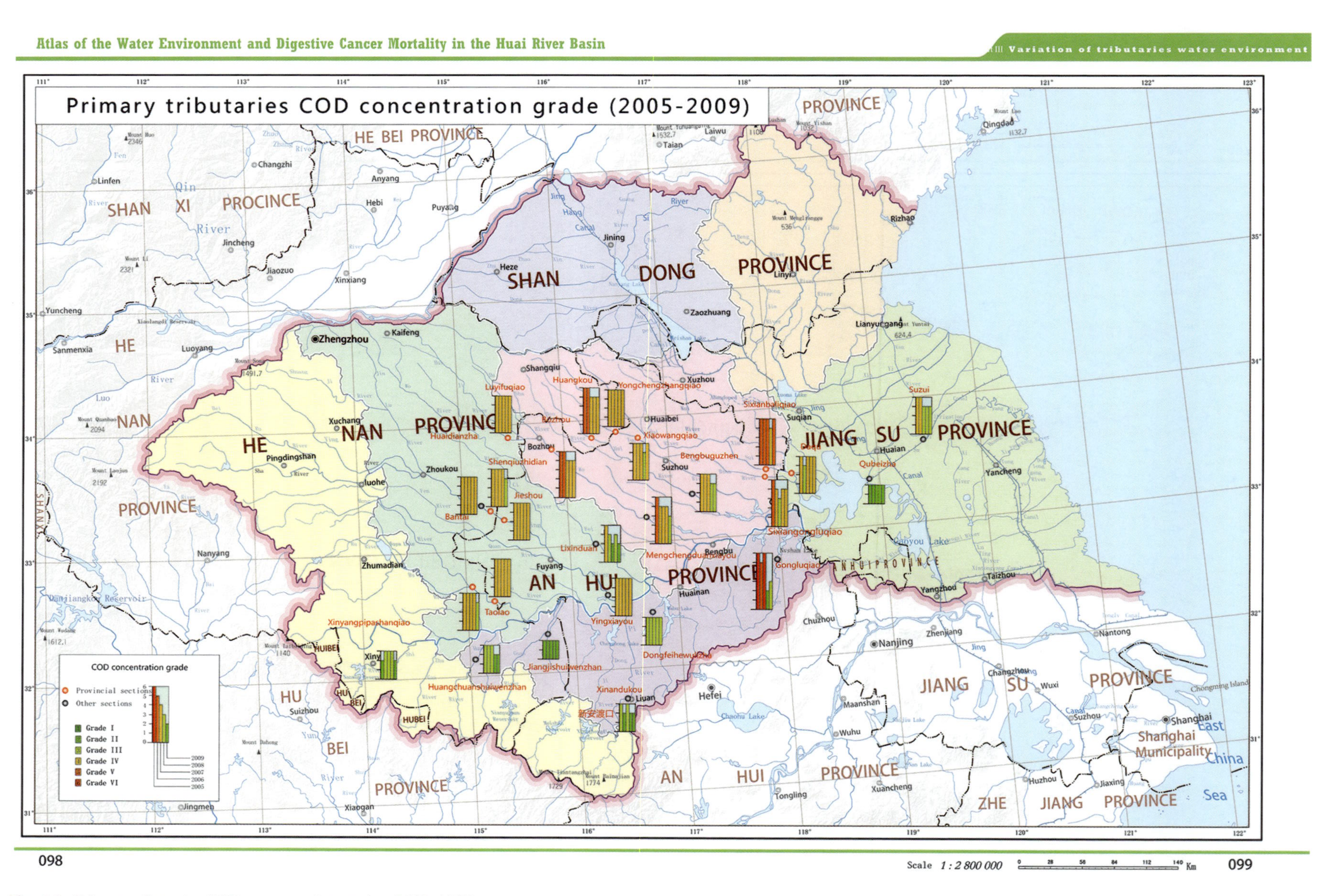

Fig. 4.9 Primary tributaries COD concentration grades (2005–2009)

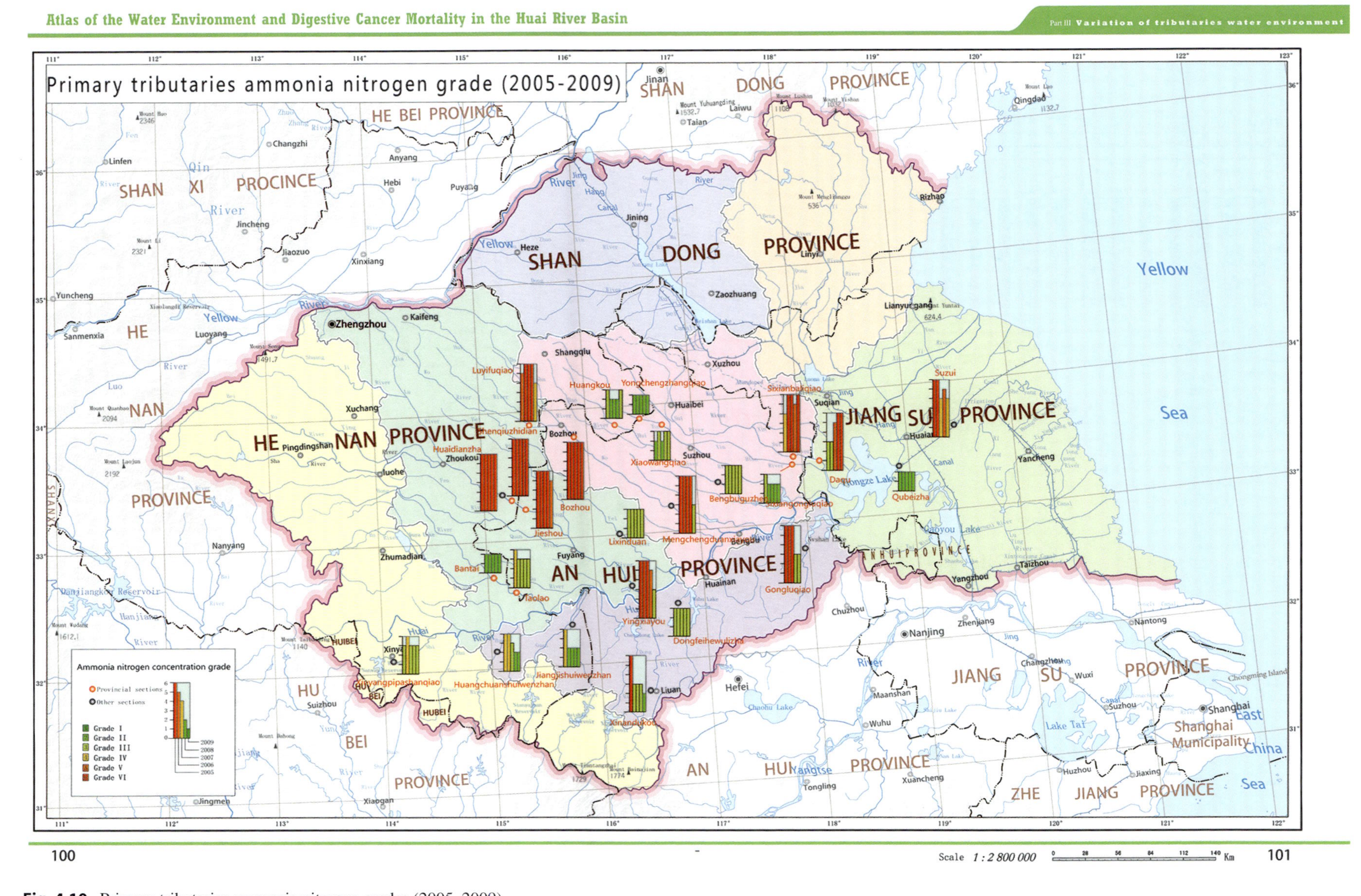

Fig. 4.10 Primary tributaries ammonia nitrogen grades (2005–2009)

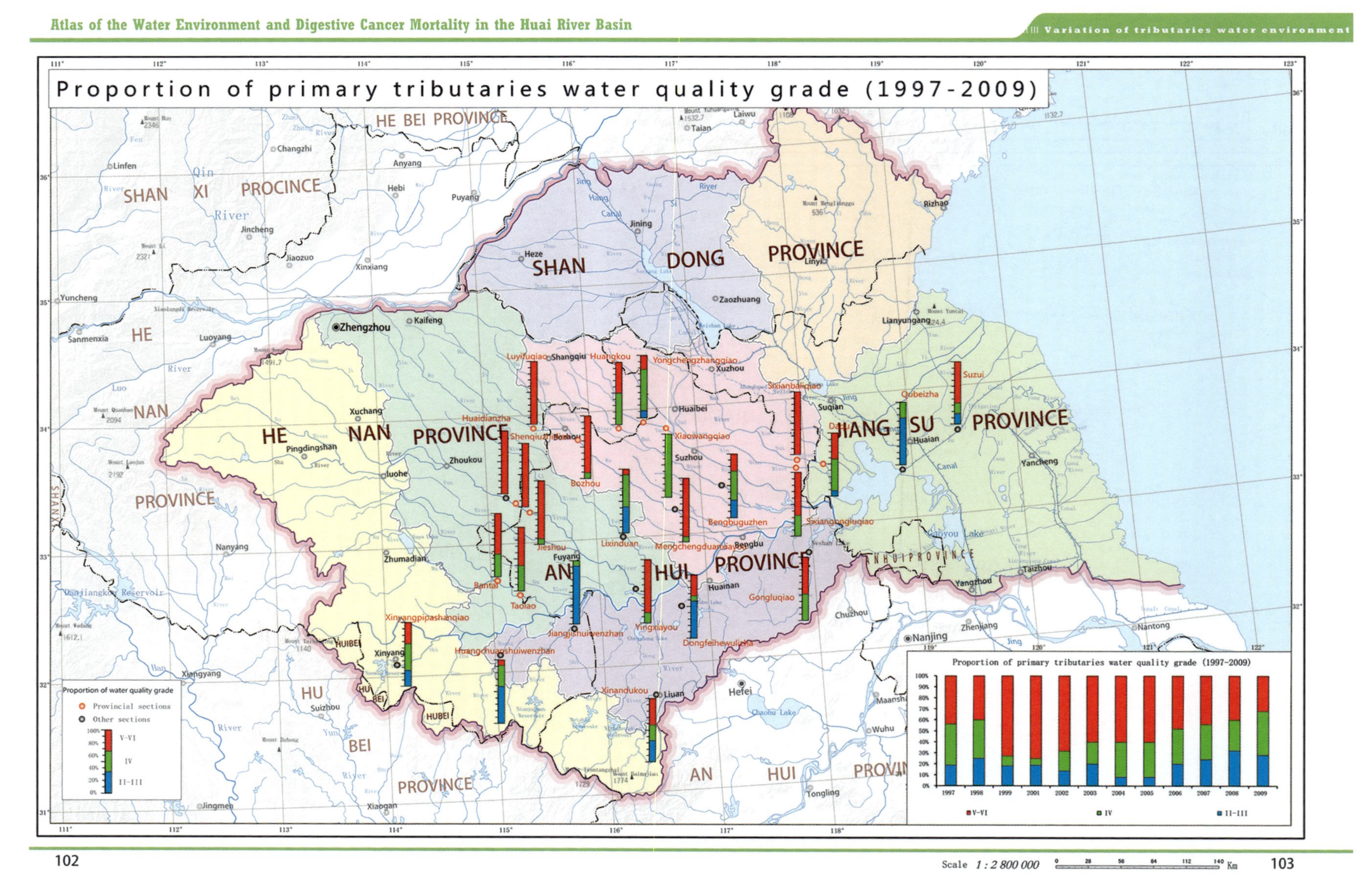

Fig. 4.11 Proportion of primary tributaries water quality grades (1997–2009)

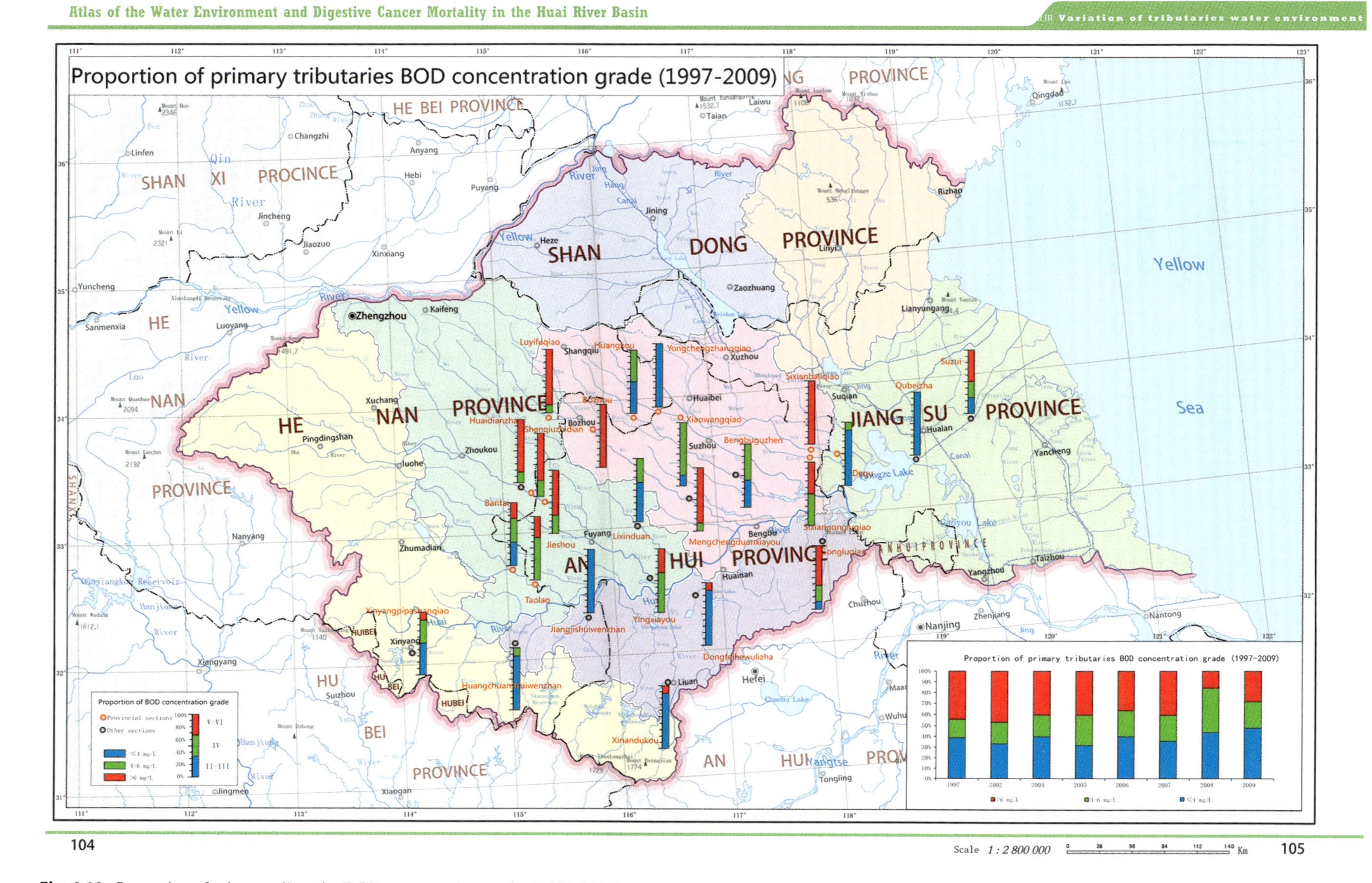

Fig. 4.12 Proportion of primary tributaries BOD concentration grades (1997–2009)

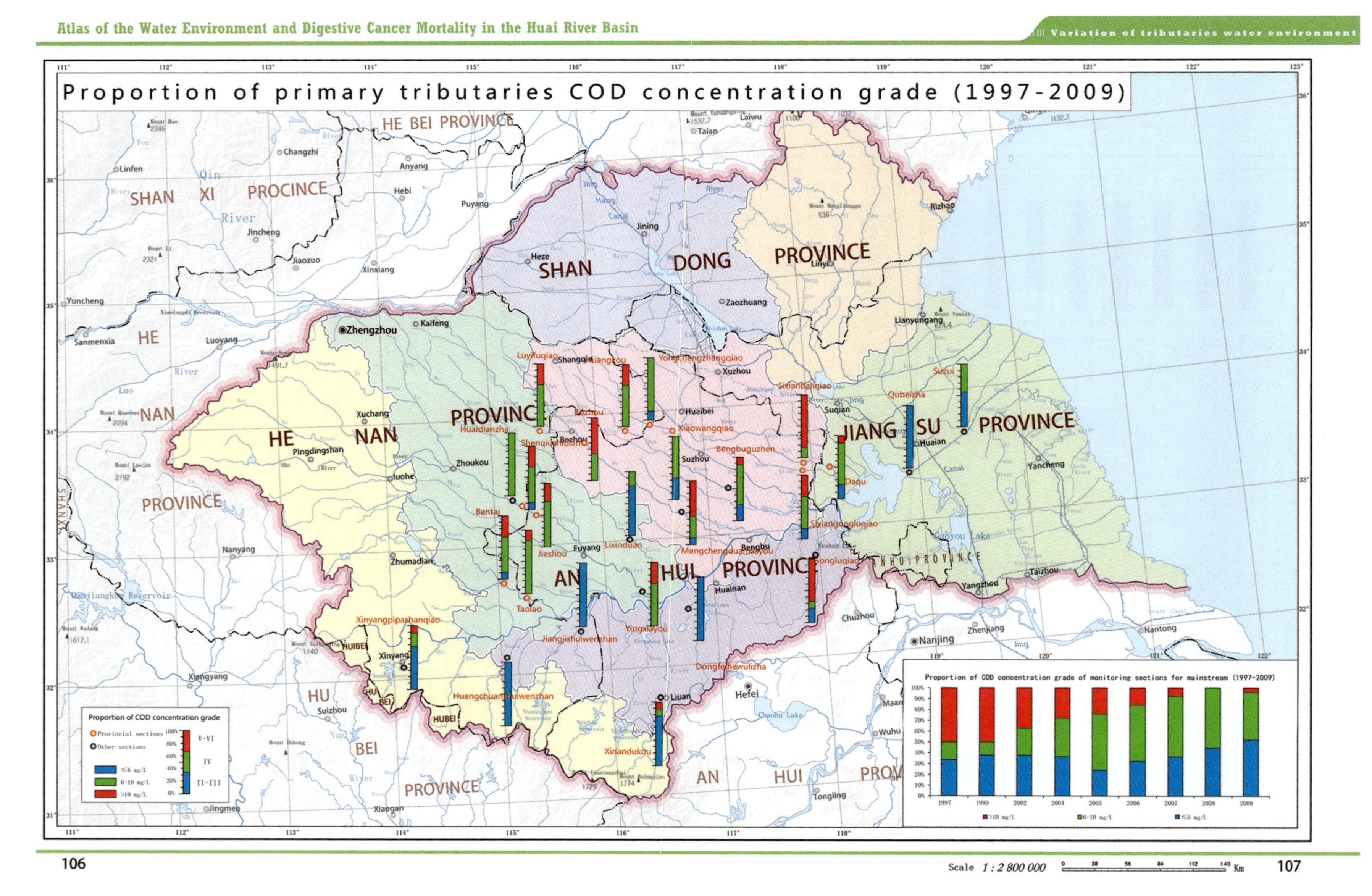

Fig. 4.13 Proportion of primary tributaries COD concentration grades (1997–2009)

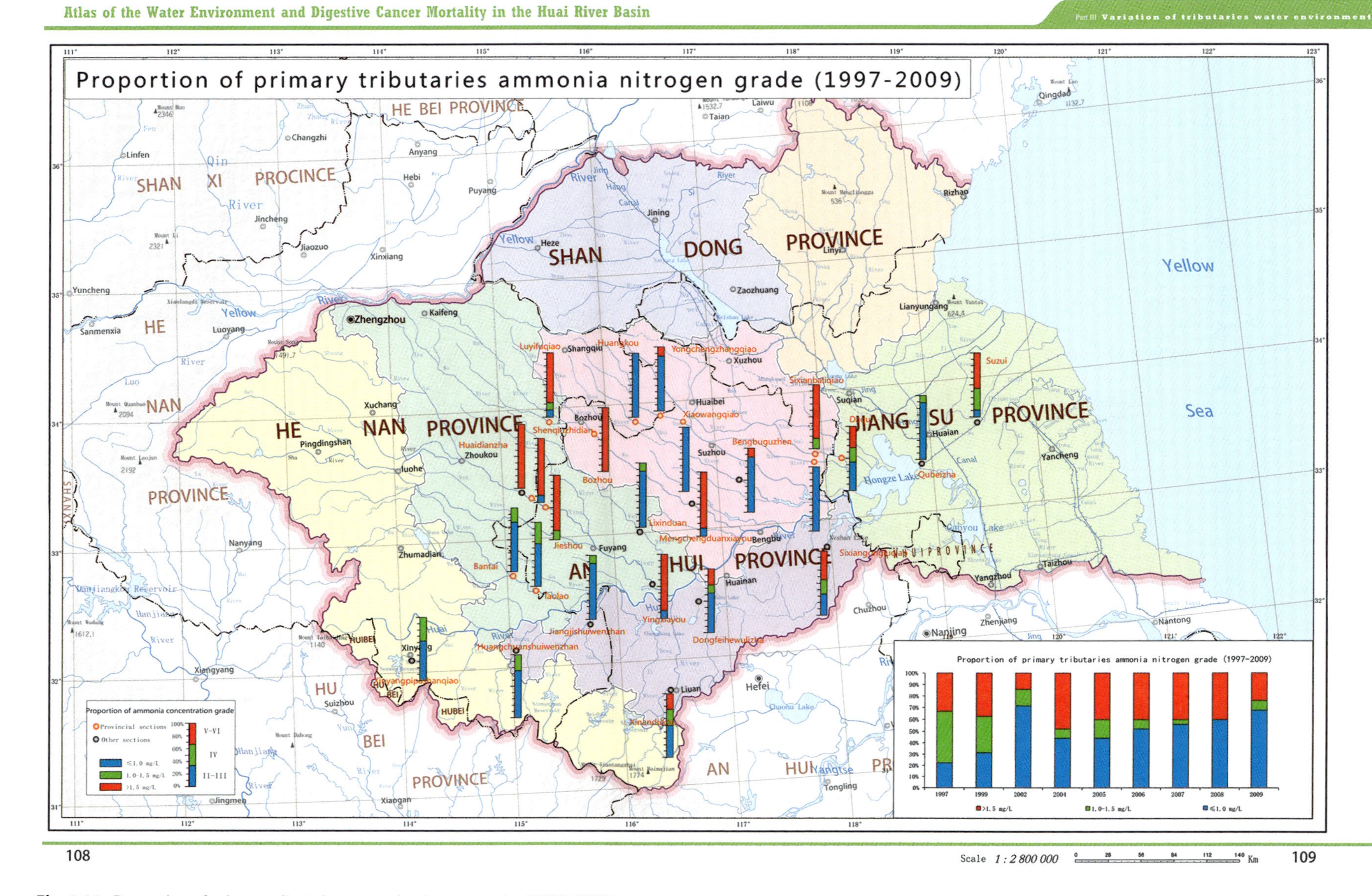

Fig. 4.14 Proportion of primary tributaries ammonia nitrogen grades (1997–2009)

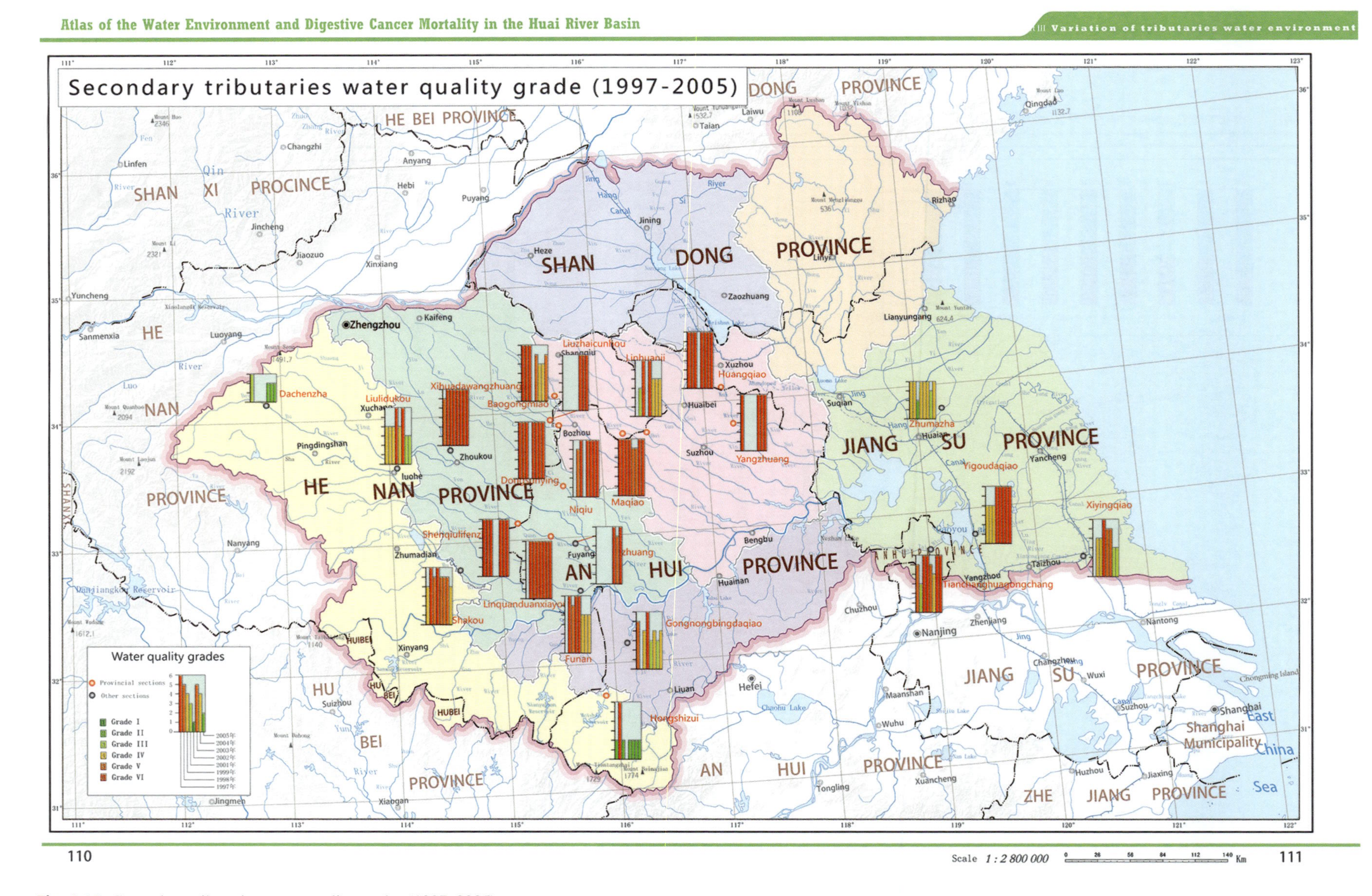

Fig. 4.15 Secondary tributaries water quality grades (1997–2005)

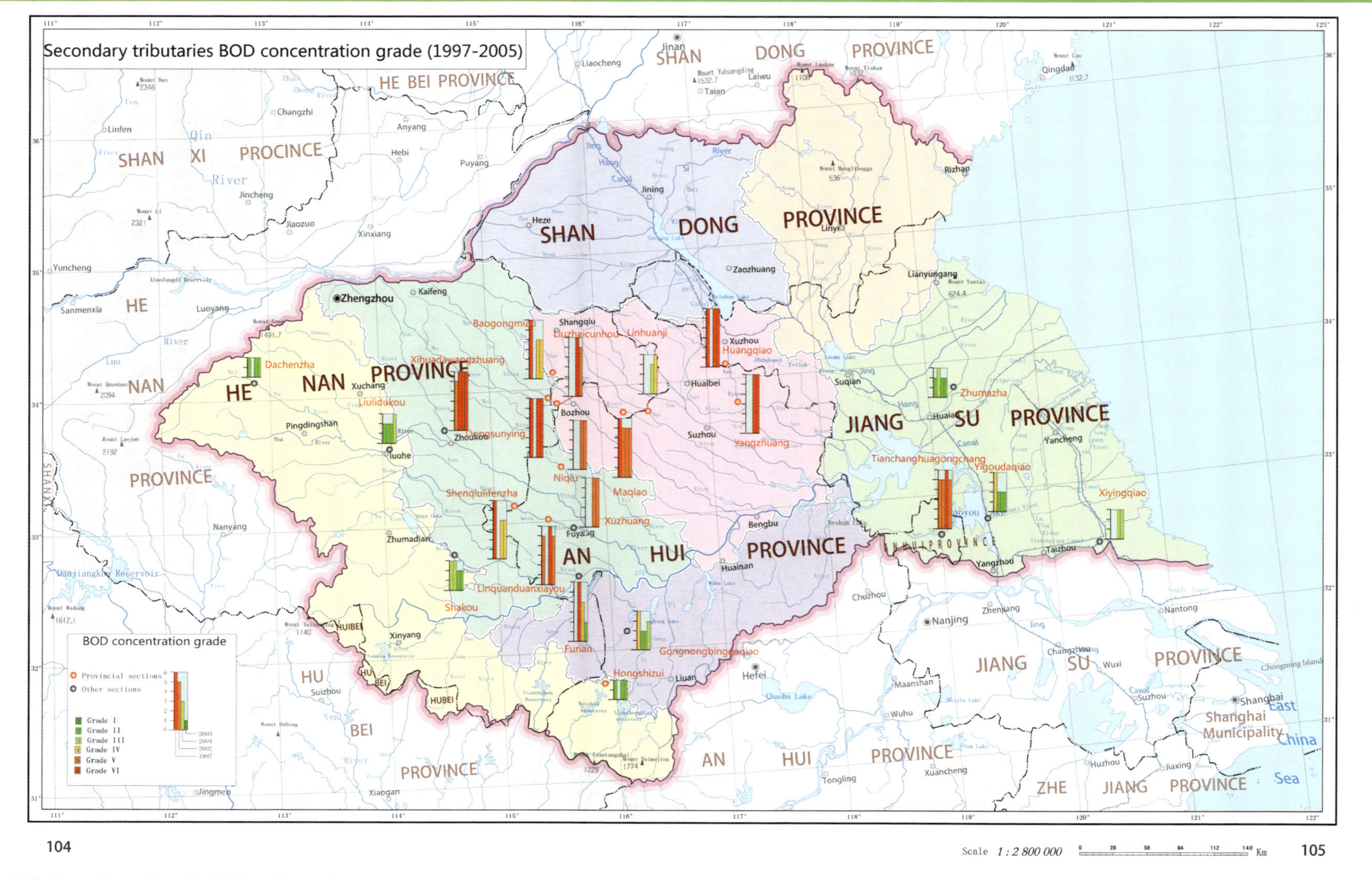

Fig. 4.16 Secondary tributaries BOD concentration grades (1997–2005)

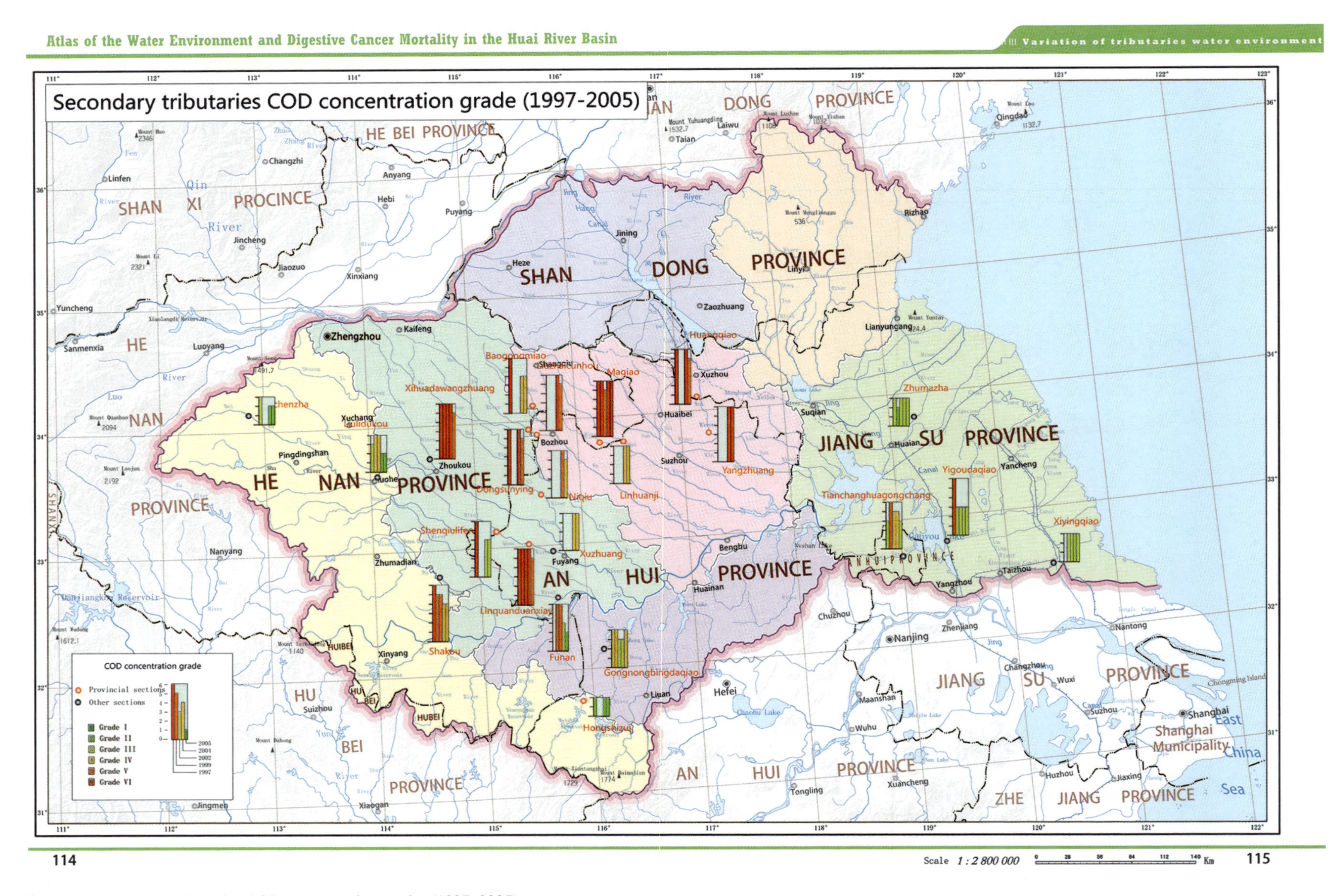

Fig. 4.17 Secondary tributaries COD concentration grades (1997–2005)

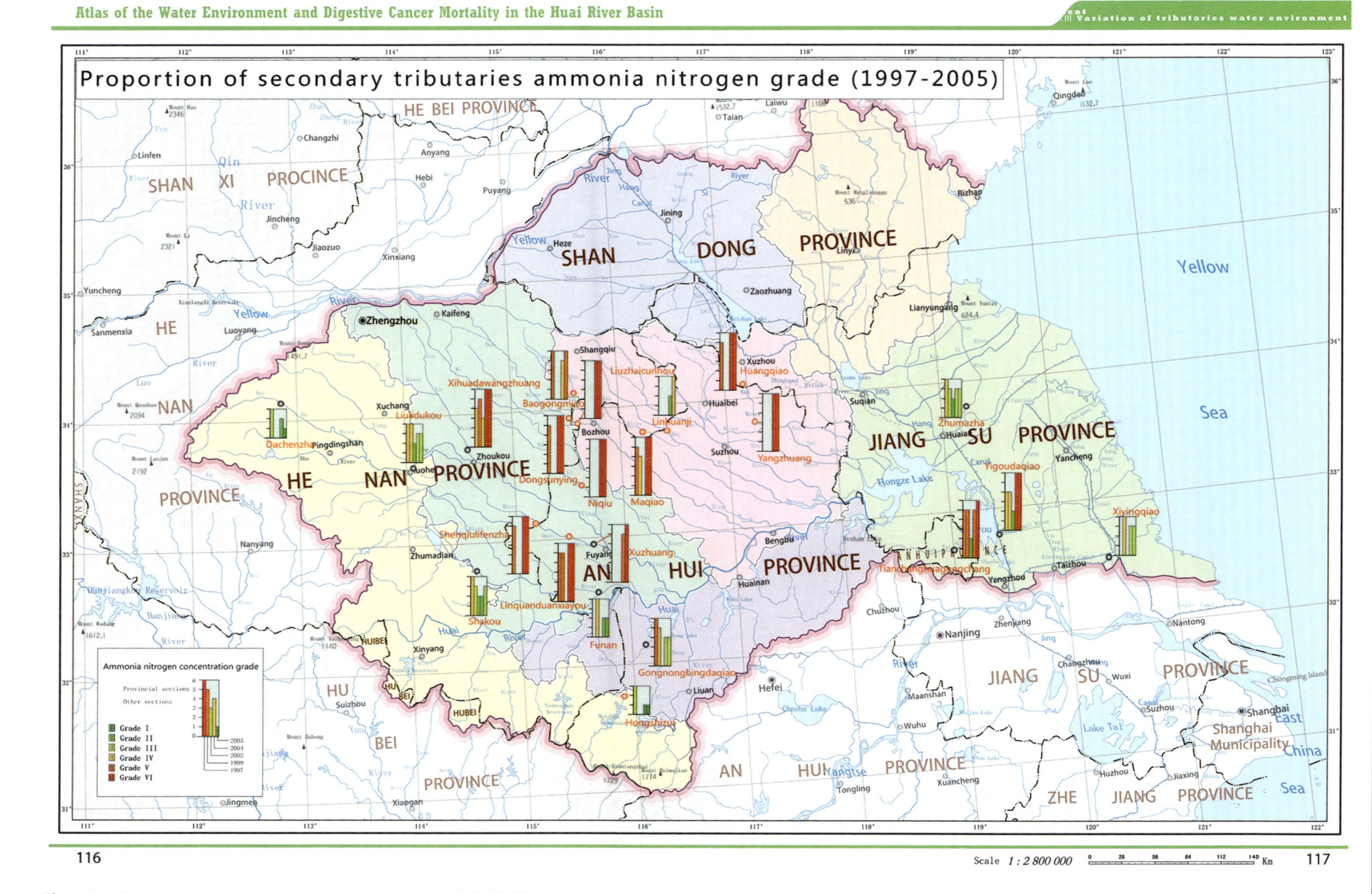

Fig. 4.18 Secondary tributaries ammonia nitrogen grades (1997–2005)

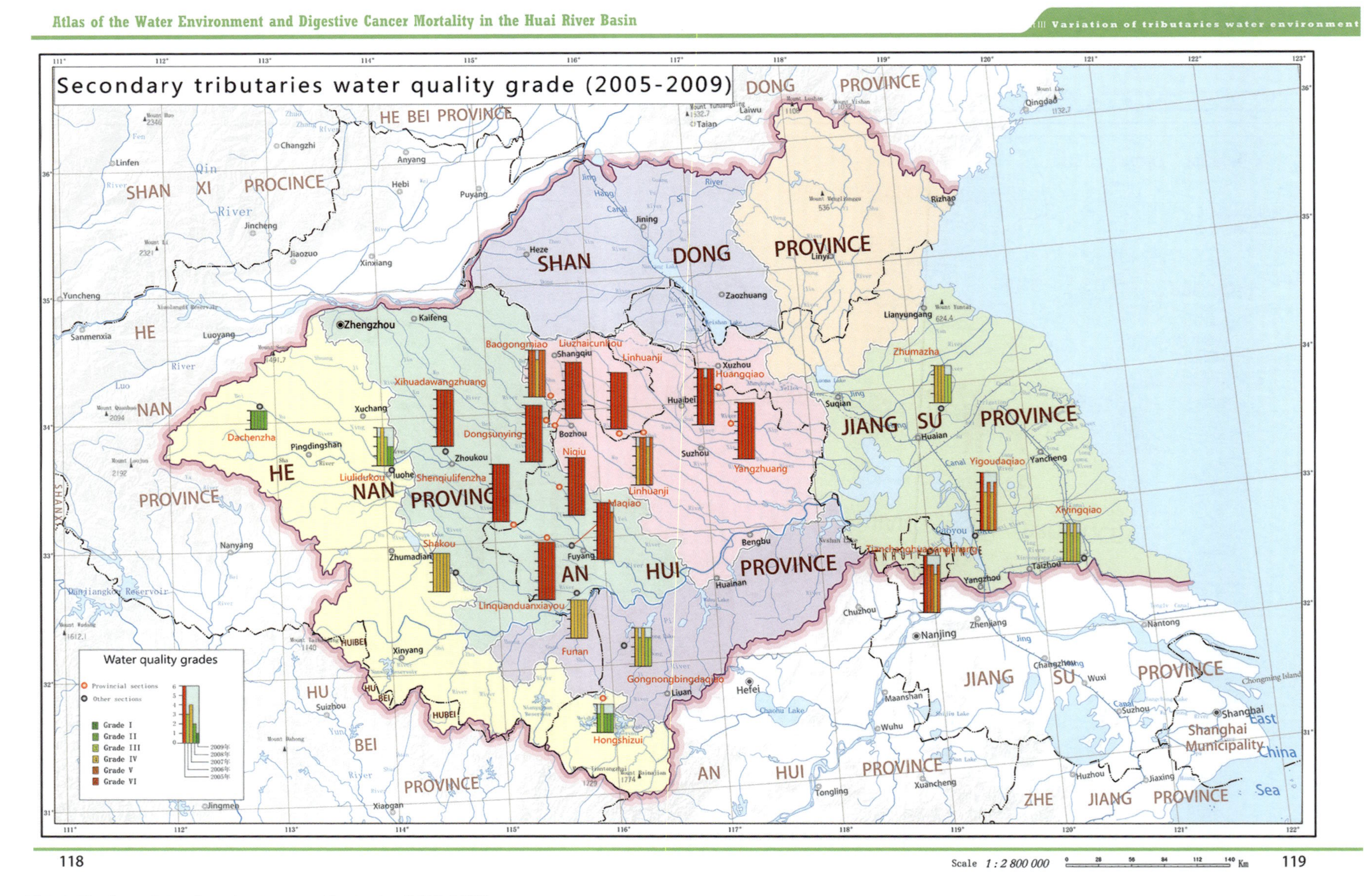

Fig. 4.19 Secondary tributaries water quality grades (2005–2009)

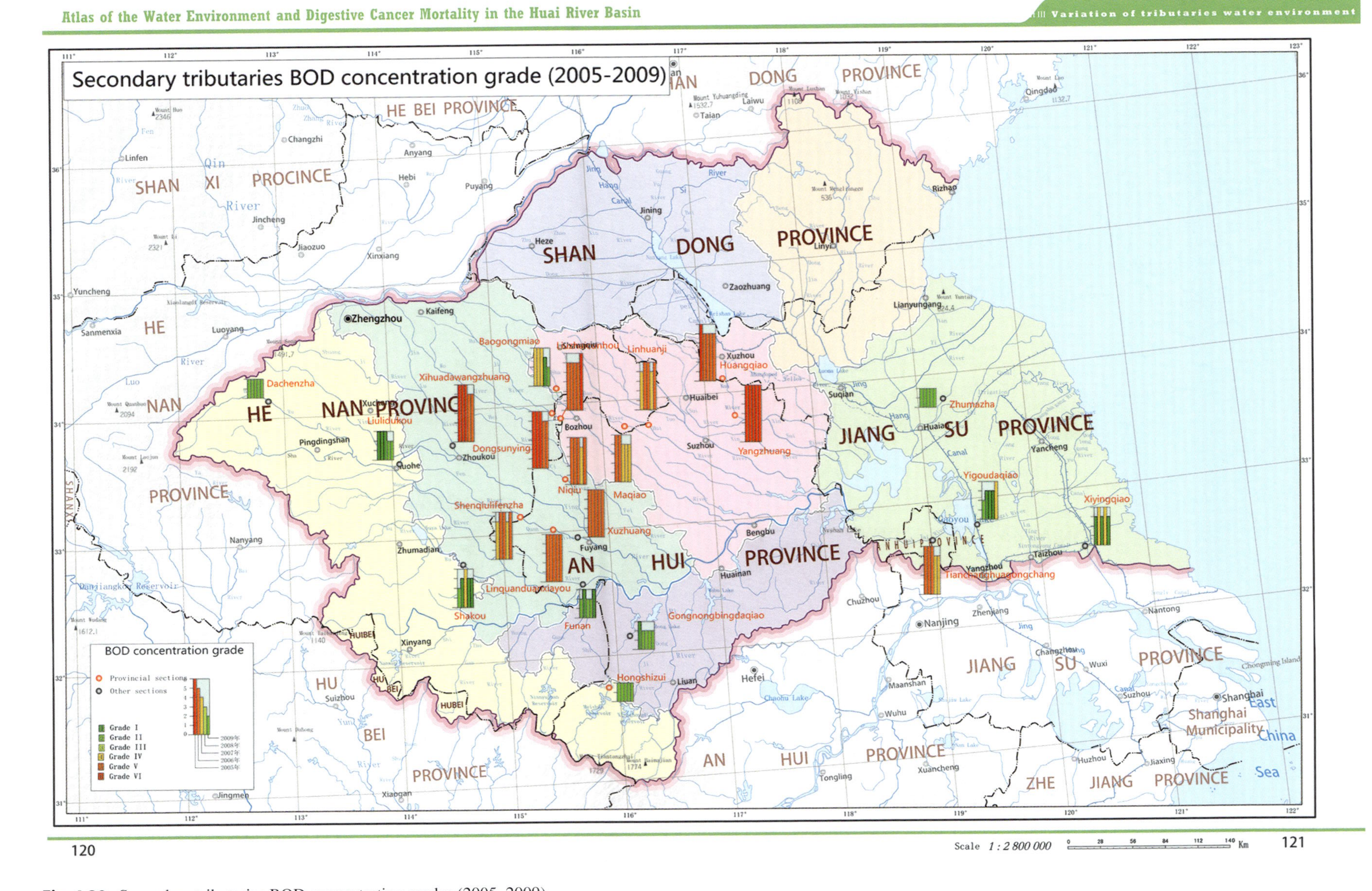

Fig. 4.20 Secondary tributaries BOD concentration grades (2005–2009)

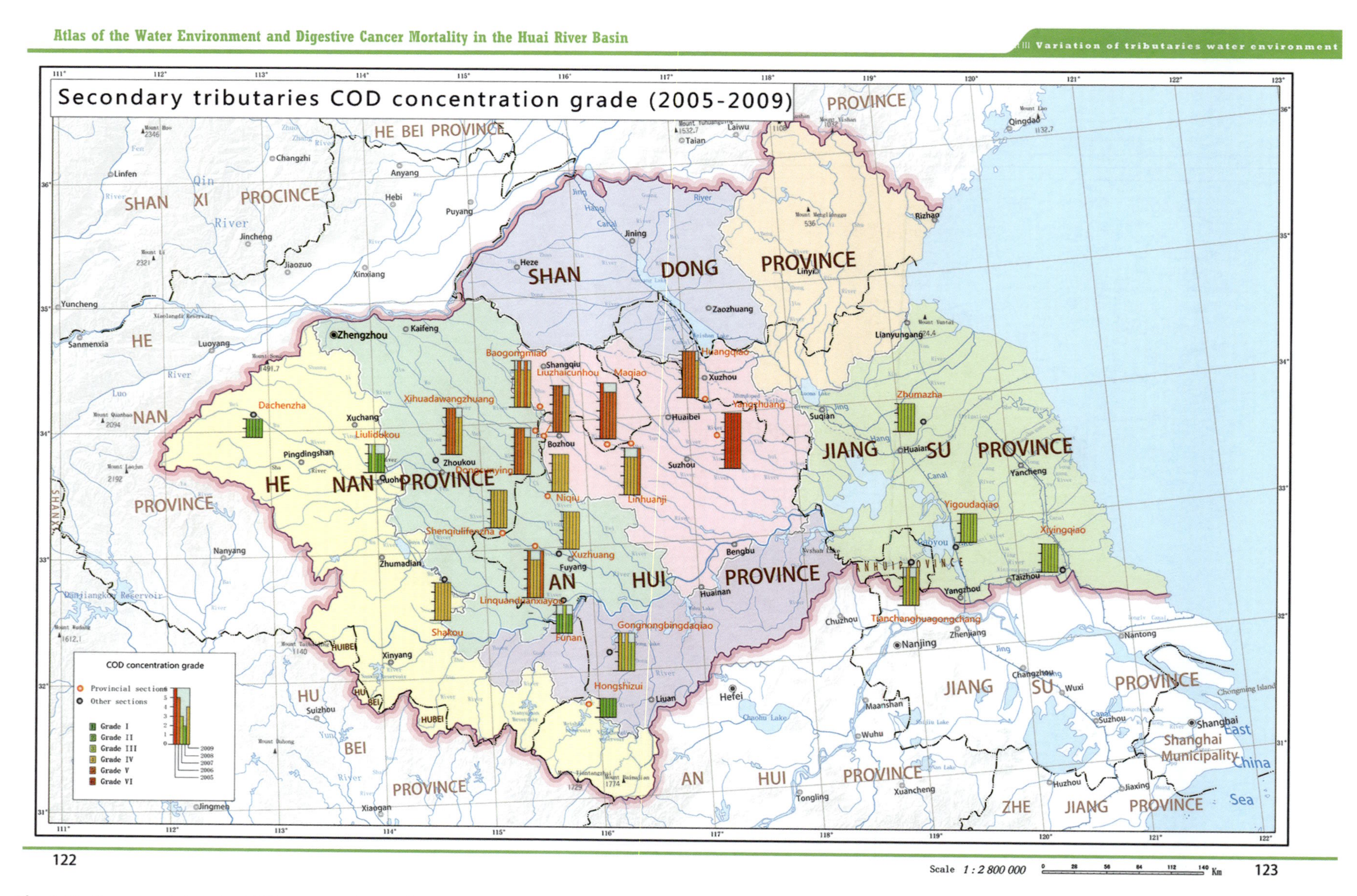

Fig. 4.21 Secondary tributaries COD concentration grades (2005–2009)

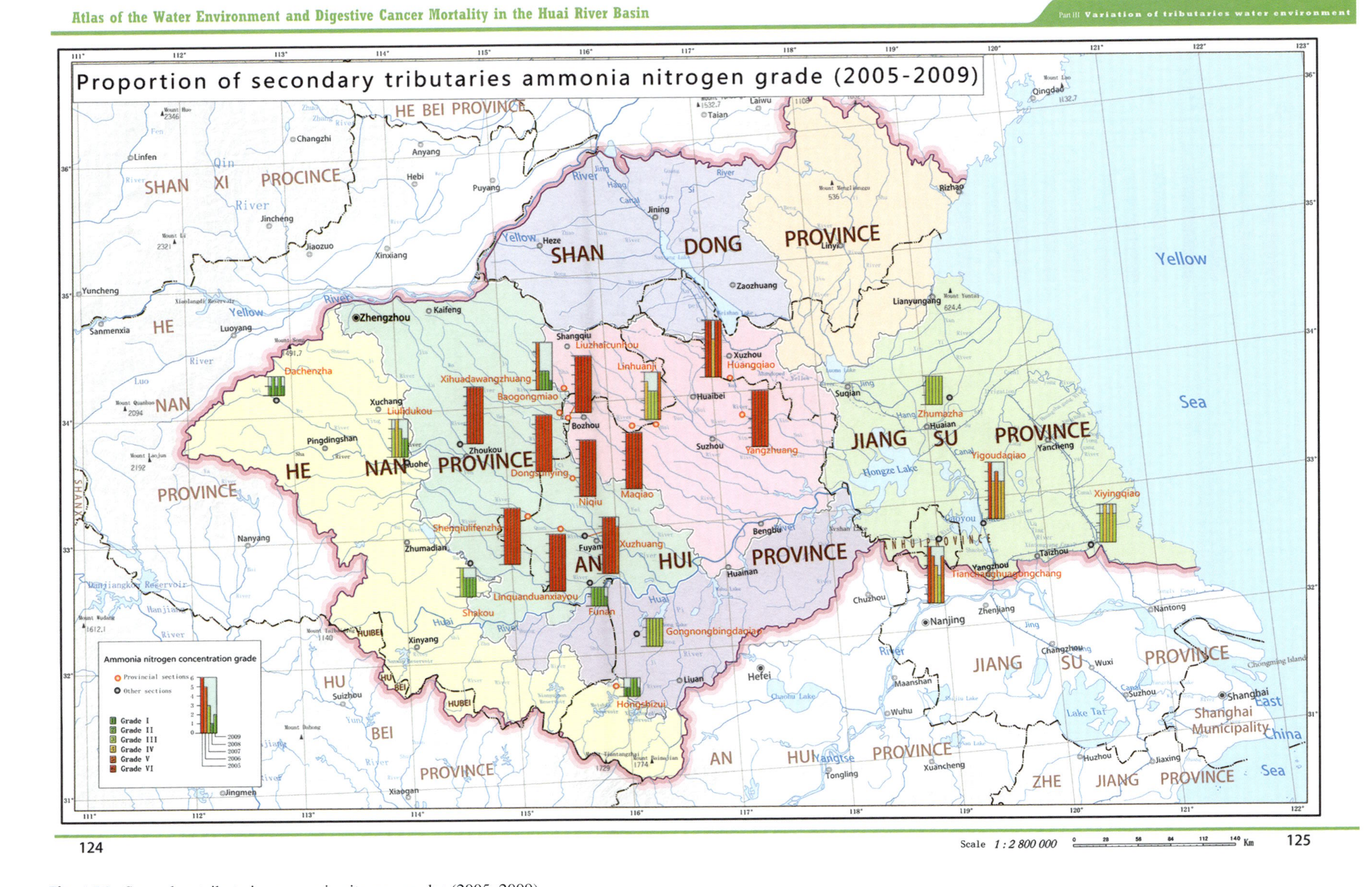

Fig. 4.22 Secondary tributaries ammonia nitrogen grades (2005–2009)

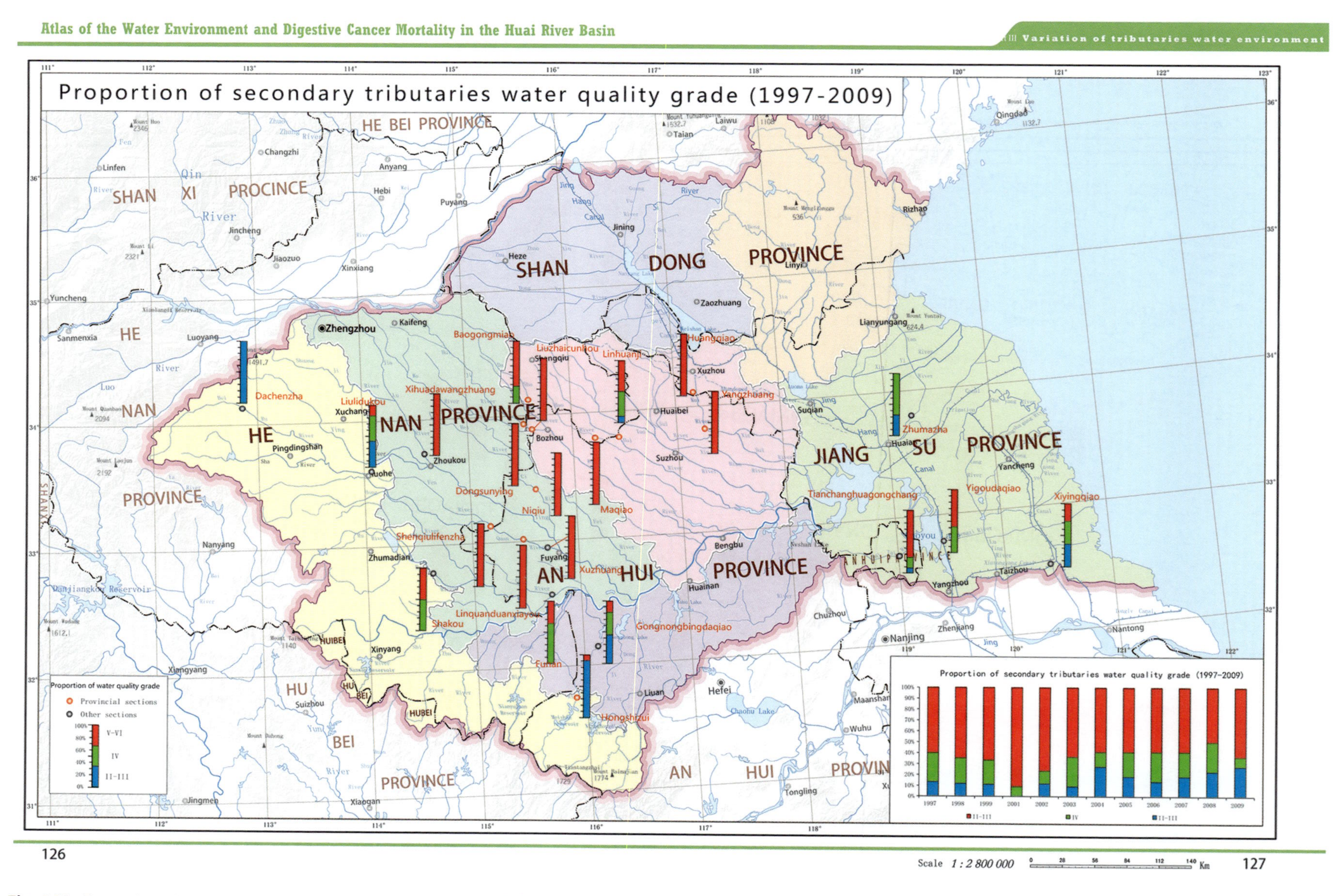

Fig. 4.23 Proportion of secondary tributaries water quality grades (1997–2009)

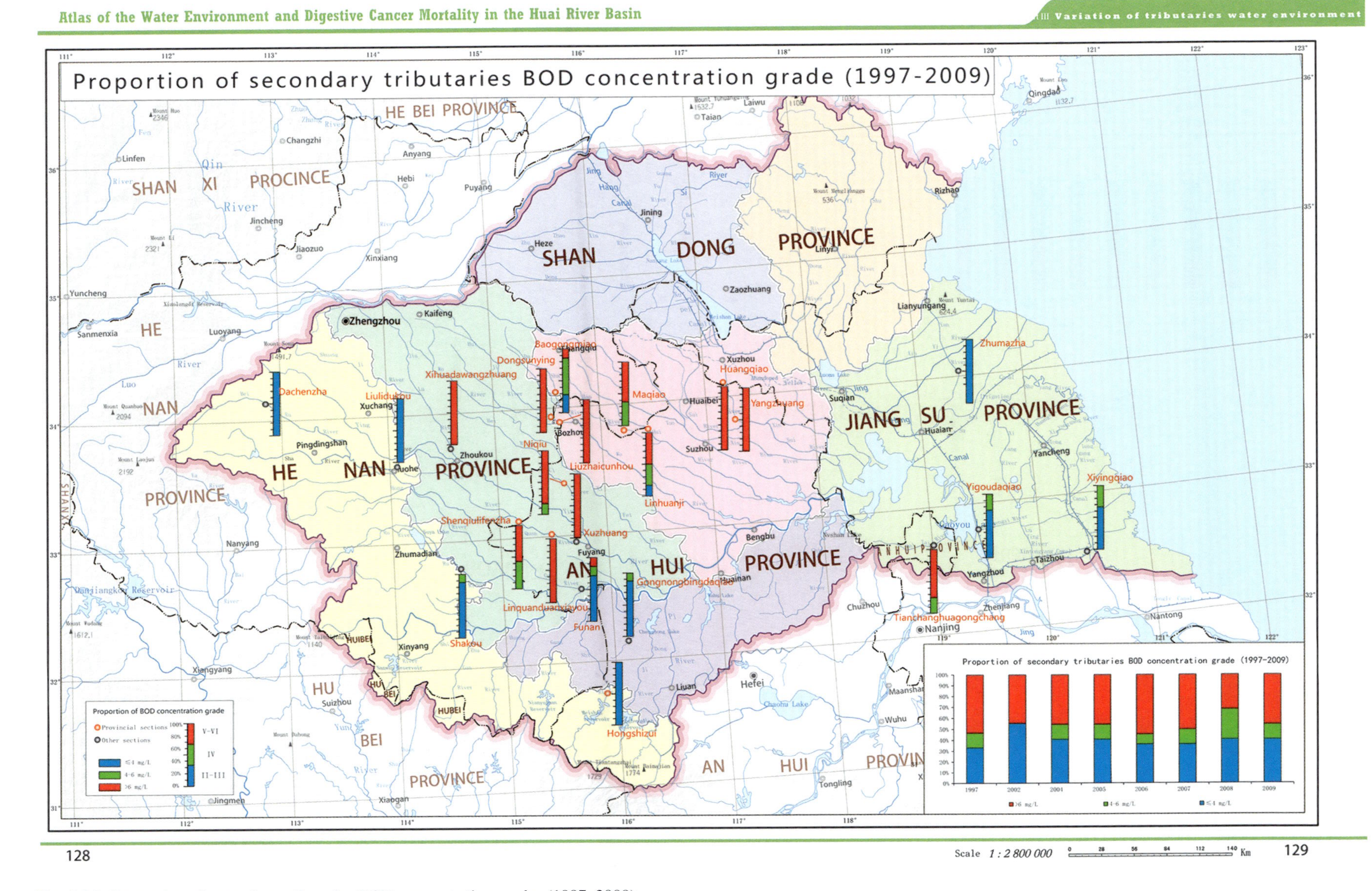

Fig. 4.24 Proportion of secondary tributaries BOD concentration grades (1997–2009)

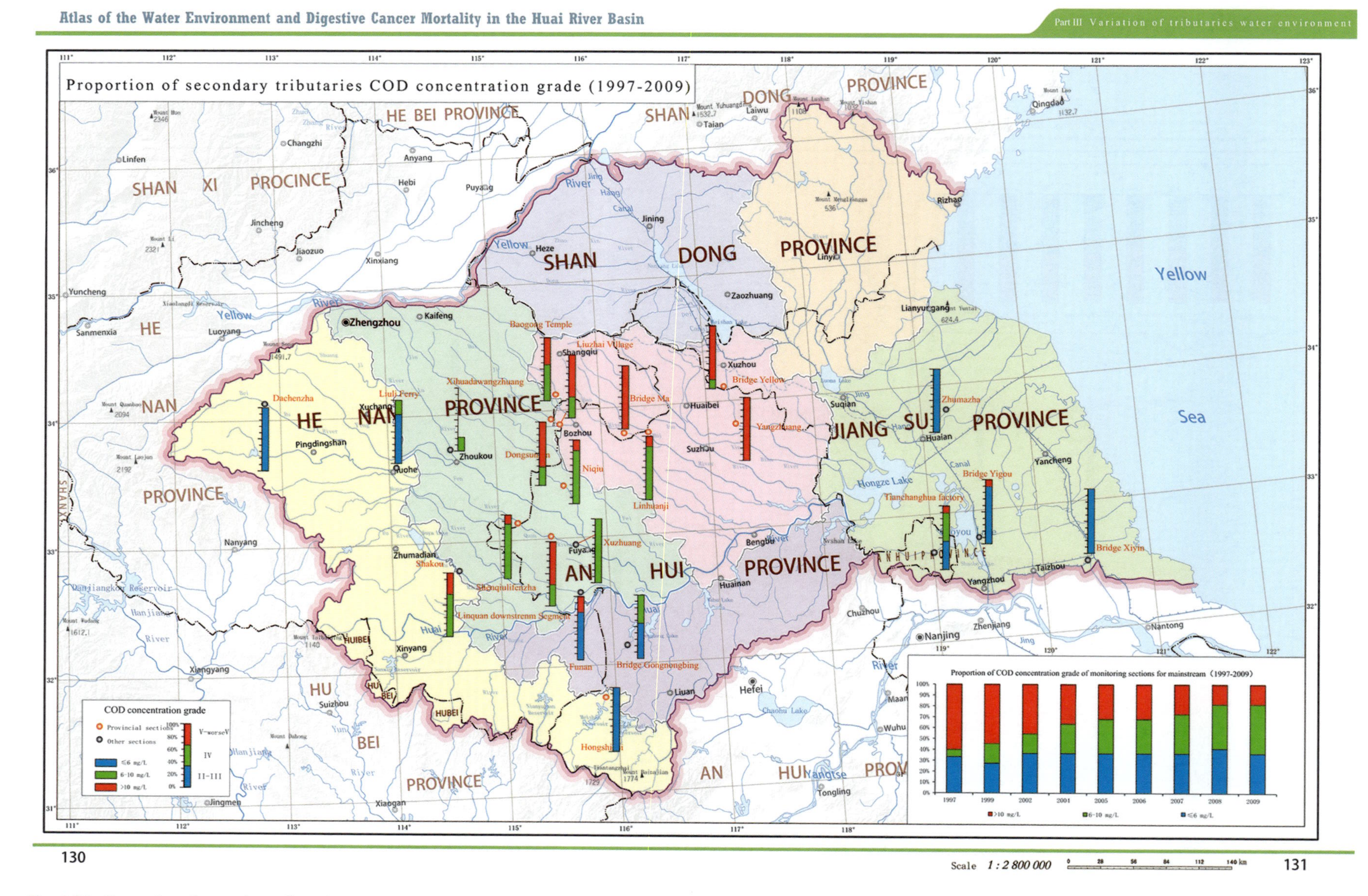

Fig. 4.25 Proportion of secondary tributaries COD concentration grades (1997–2009)

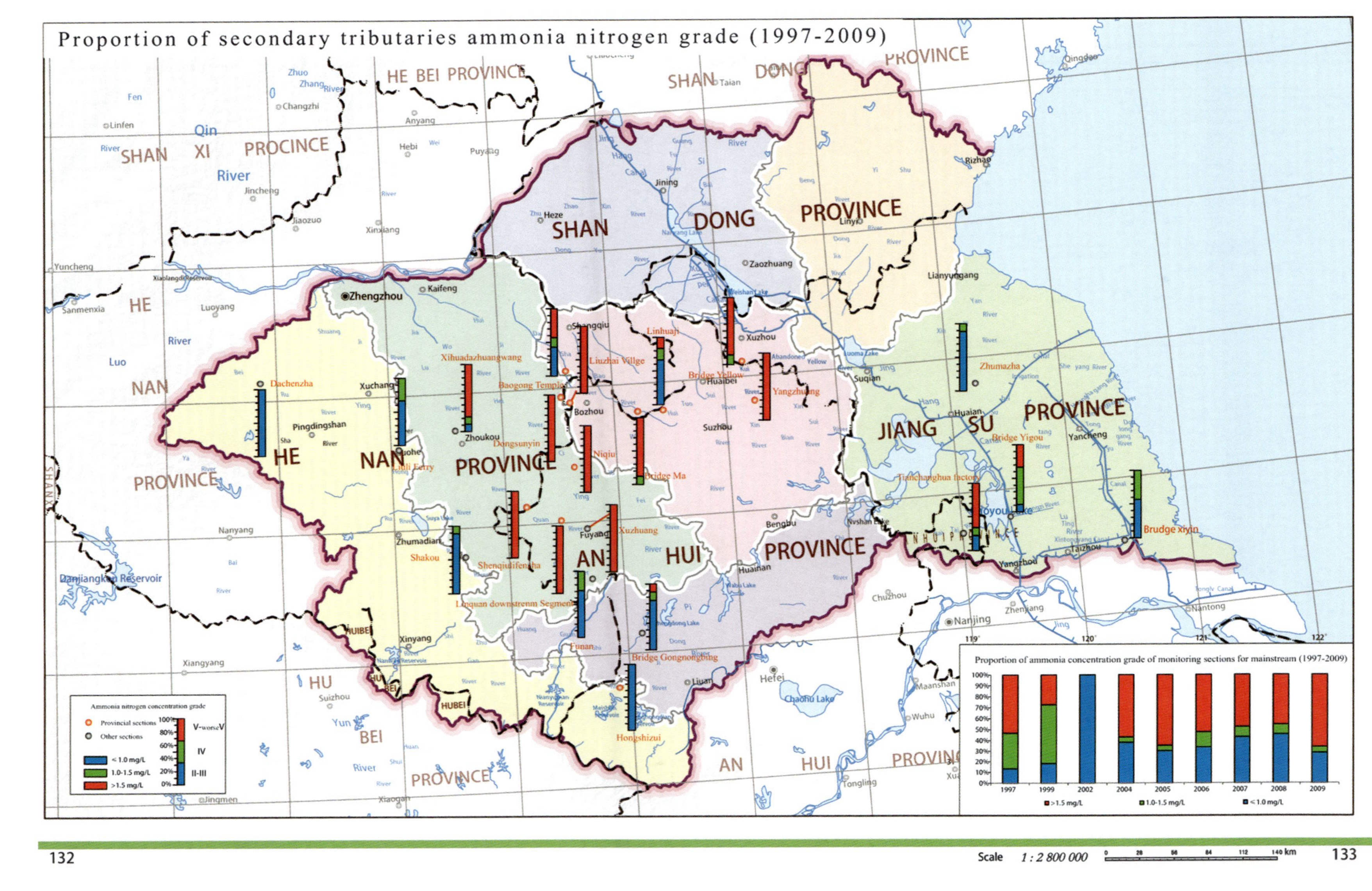

Fig. 4.26 Proportion of secondary tributaries ammonia nitrogen grades (1997–2009)

Fig. 4.27 Beijing-Hangzhou Canal water quality grades (1997–2005)

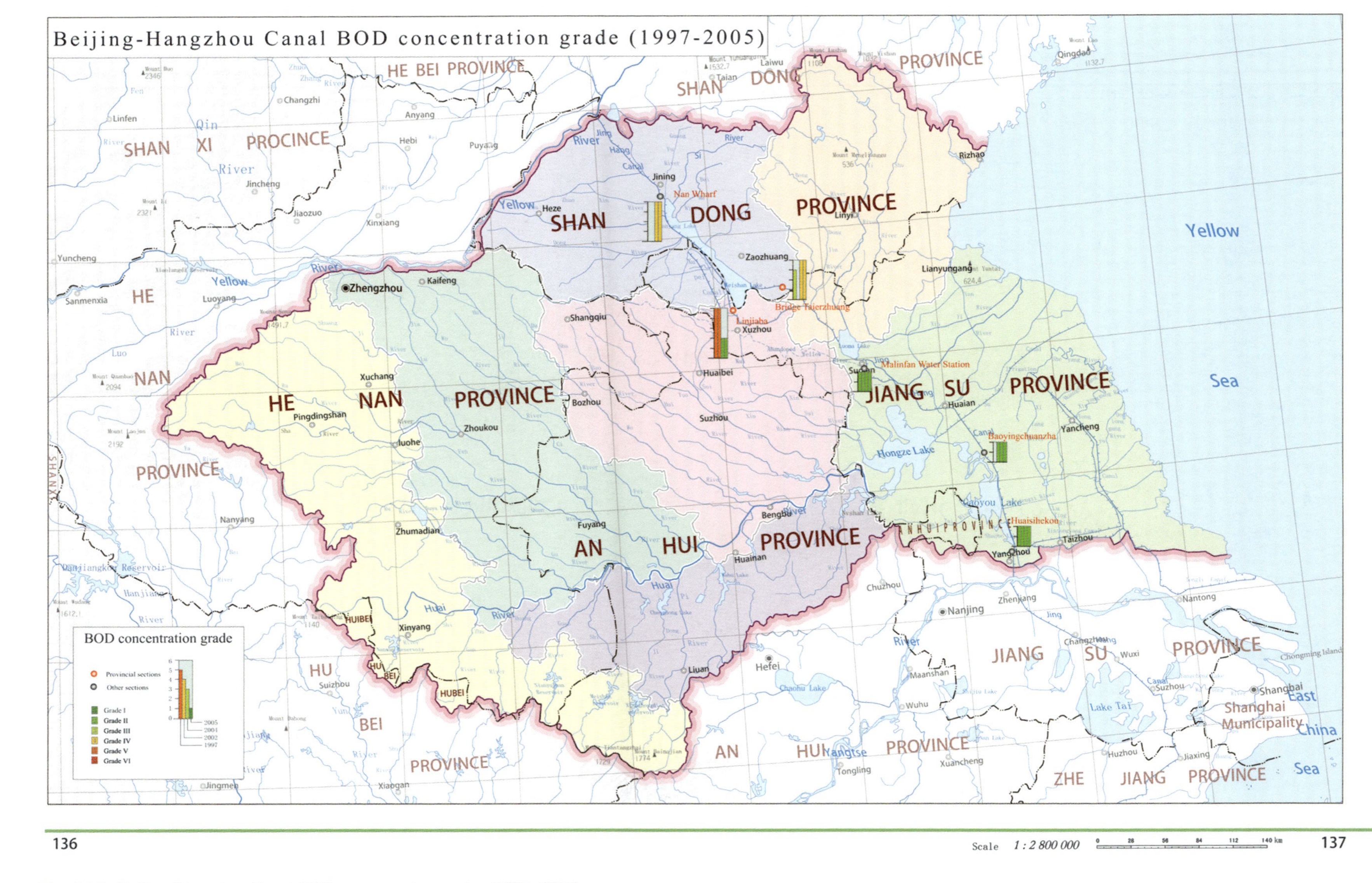

Fig. 4.28 Beijing-Hangzhou Canal BOD concentration grades (1997–2005)

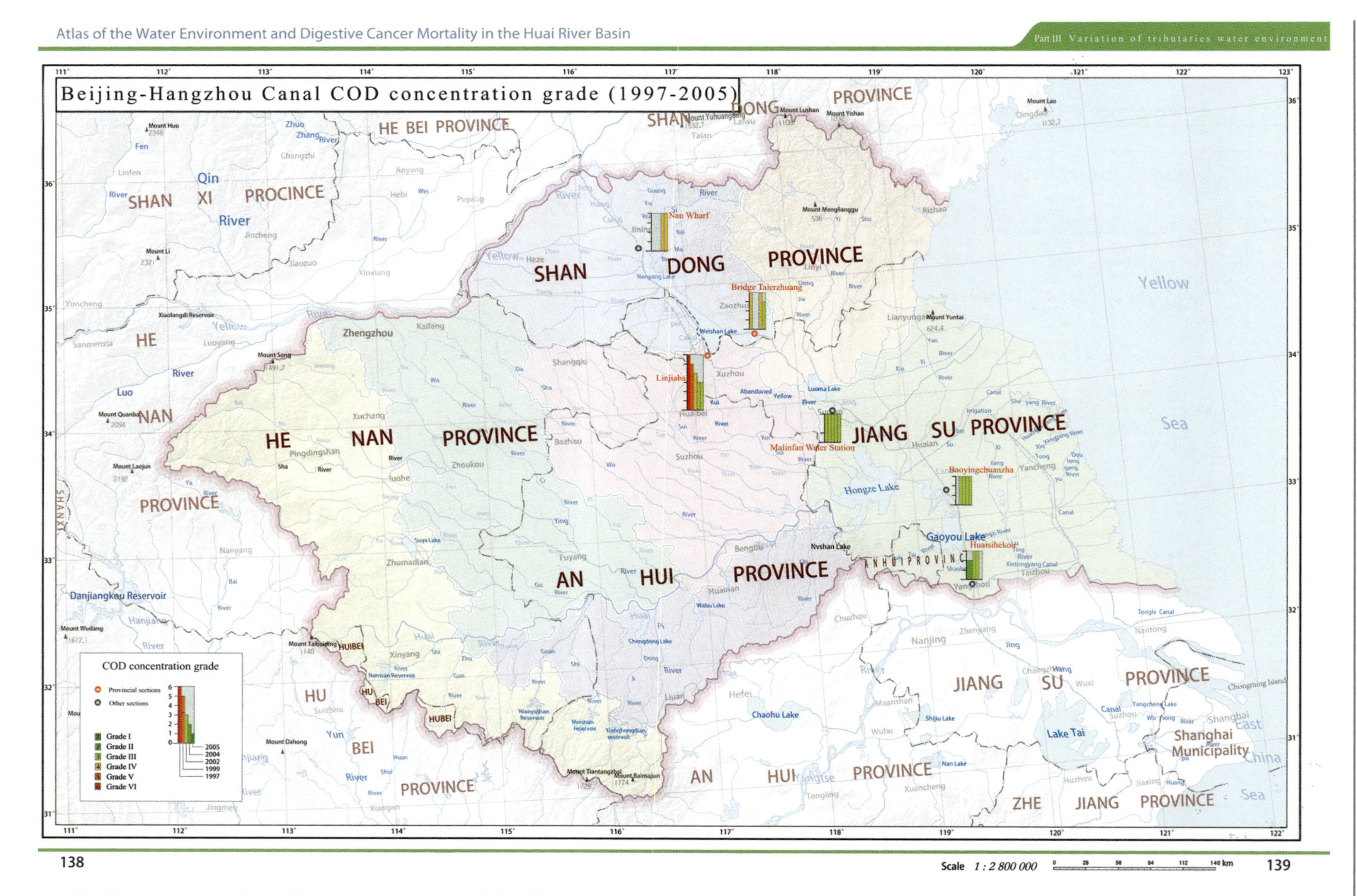

Fig. 4.29 Beijing-Hangzhou Canal COD concentration grades (1997–2005)

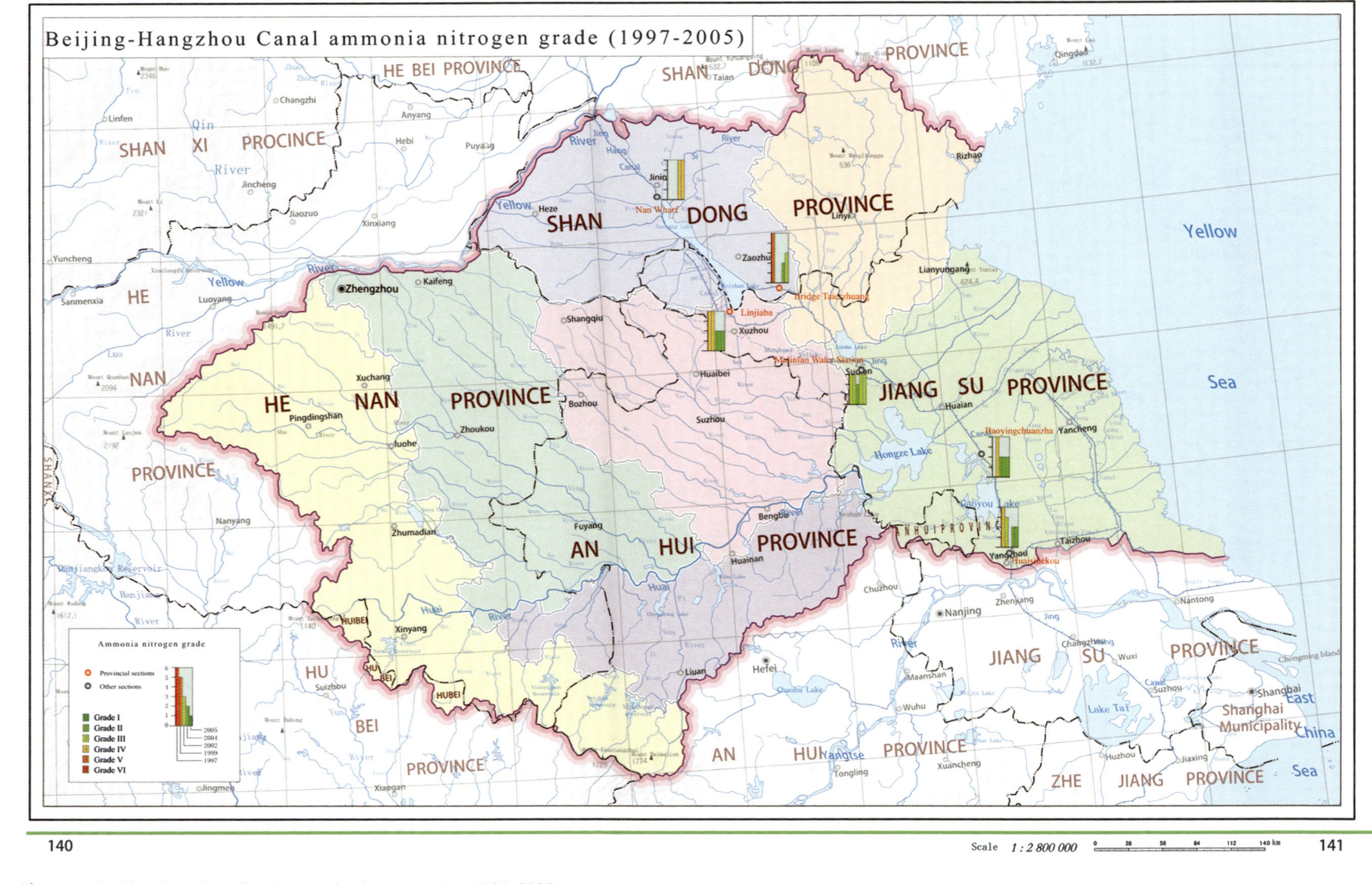

Fig. 4.30 Beijing-Hangzhou Canal ammonia nitrogen grades (1997–2005)

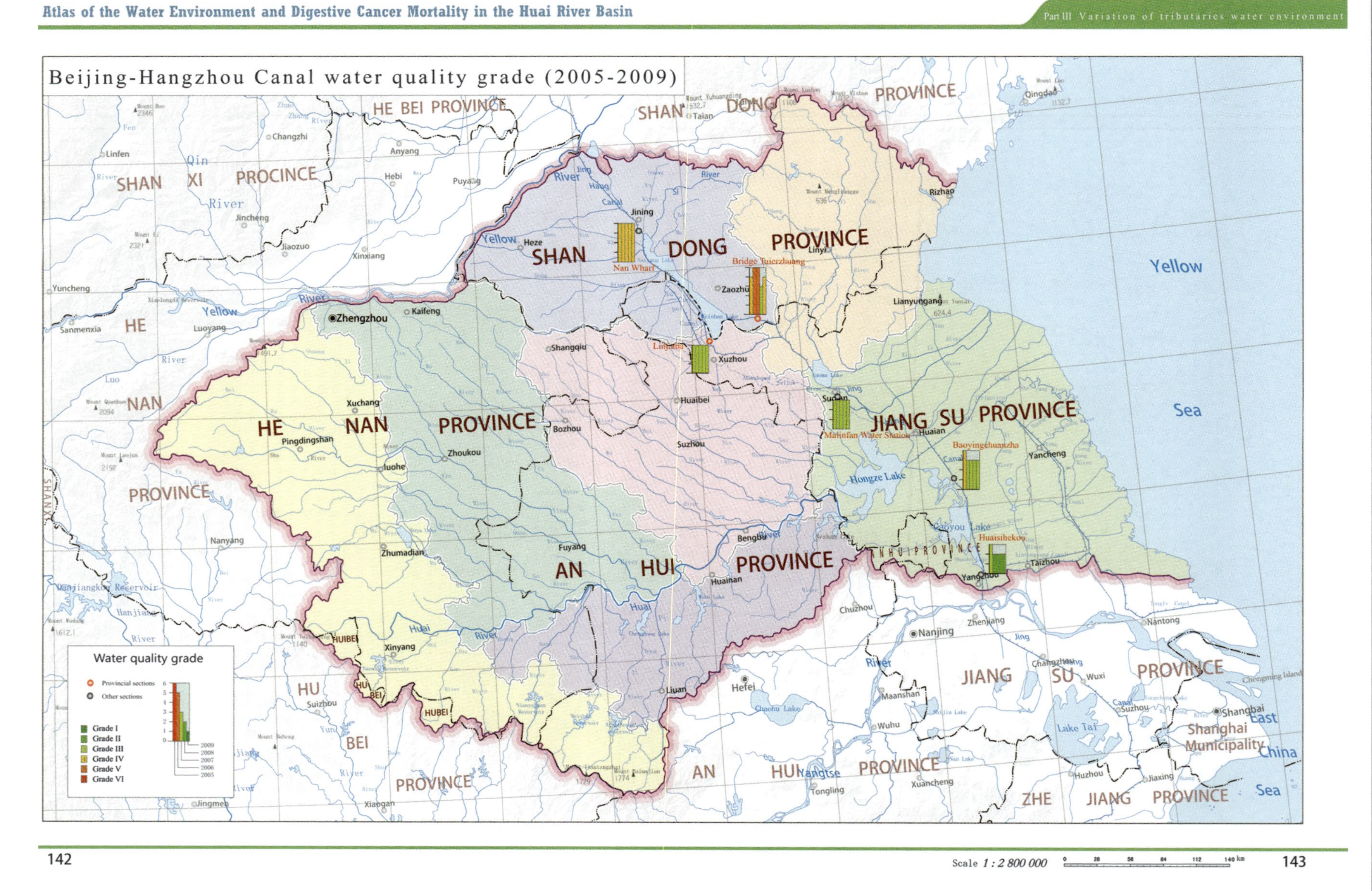

Fig. 4.31 Beijing-Hangzhou Canal water quality grades (2005–2009)

Fig. 4.32 Beijing-Hangzhou Canal BOD concentration grades (2005–2009)

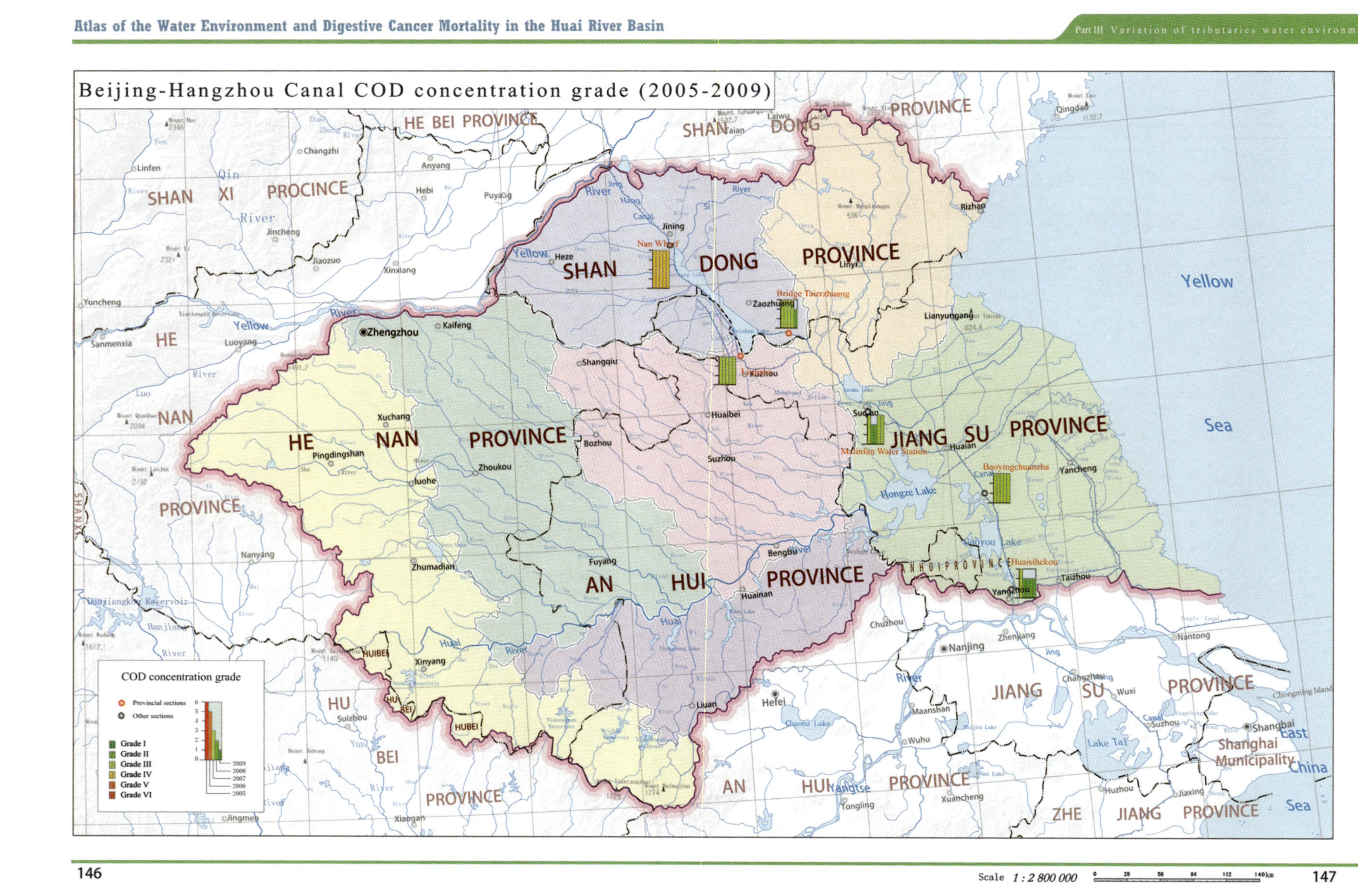

Fig. 4.33 Beijing-Hangzhou Canal COD concentration grades (2005–2009)

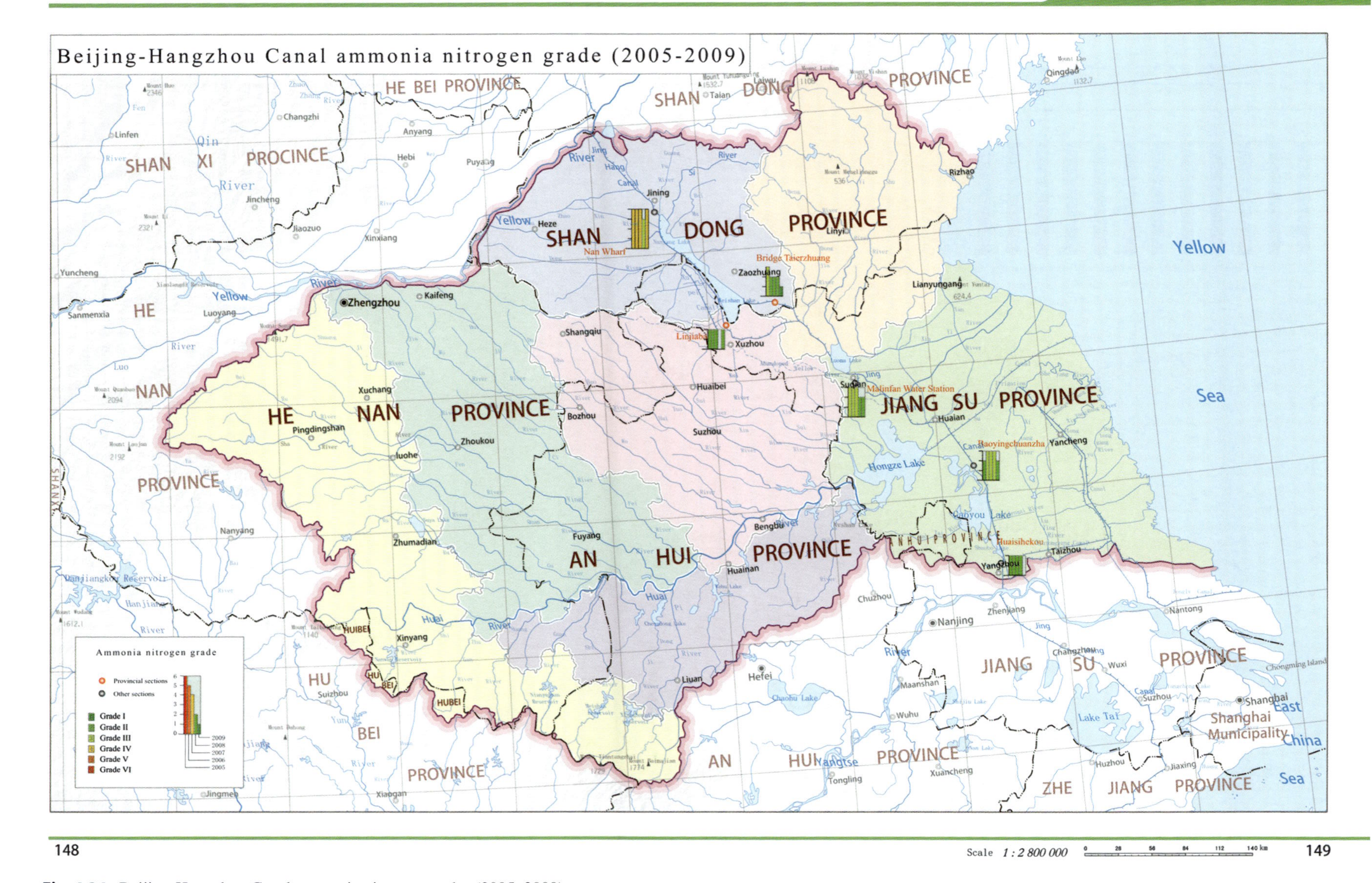

Fig. 4.34 Beijing-Hangzhou Canal ammonia nitrogen grades (2005–2009)

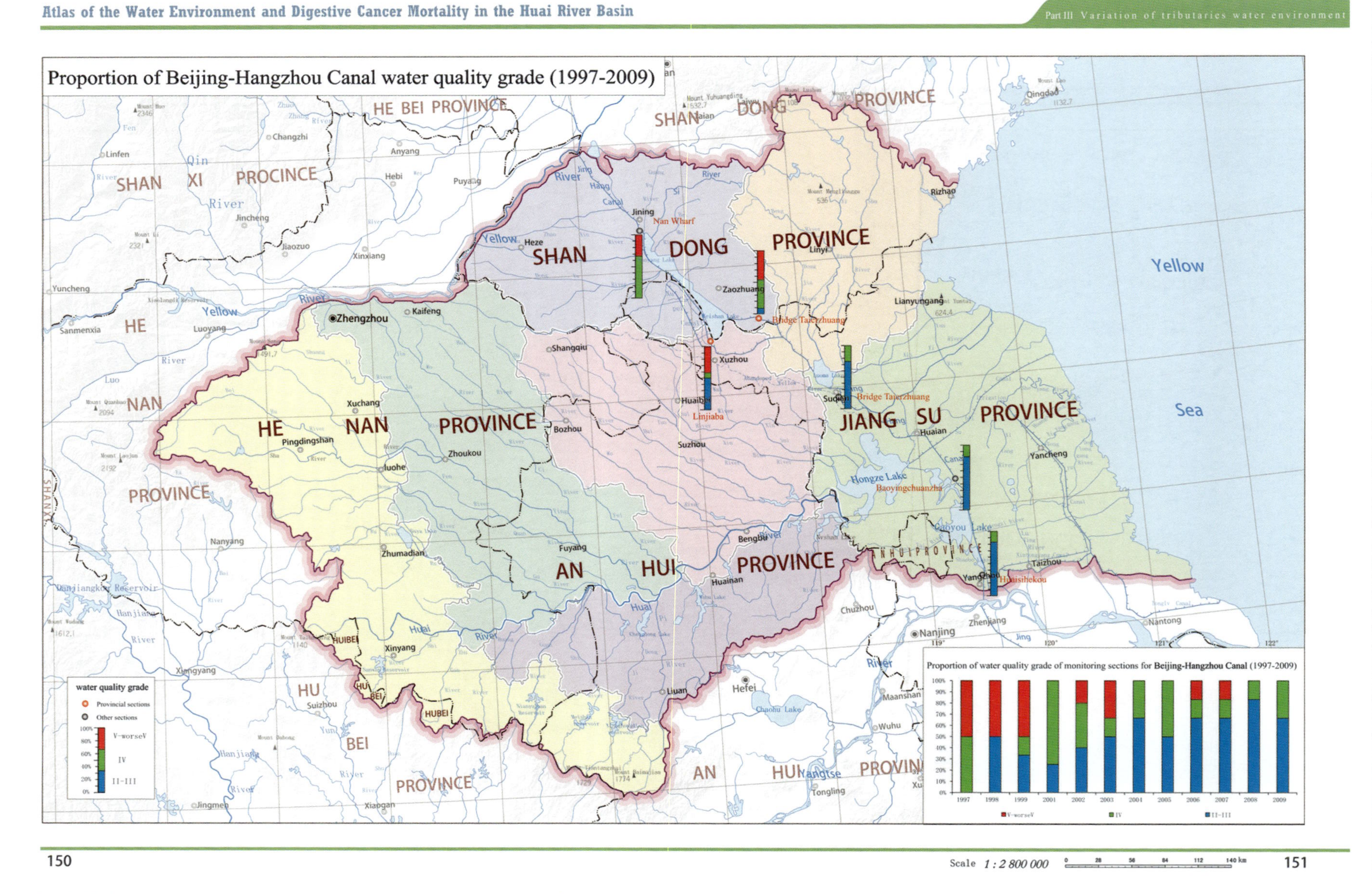

Fig. 4.35 Proportion of Beijing-Hangzhou Canal water quality grades (1997–2009)

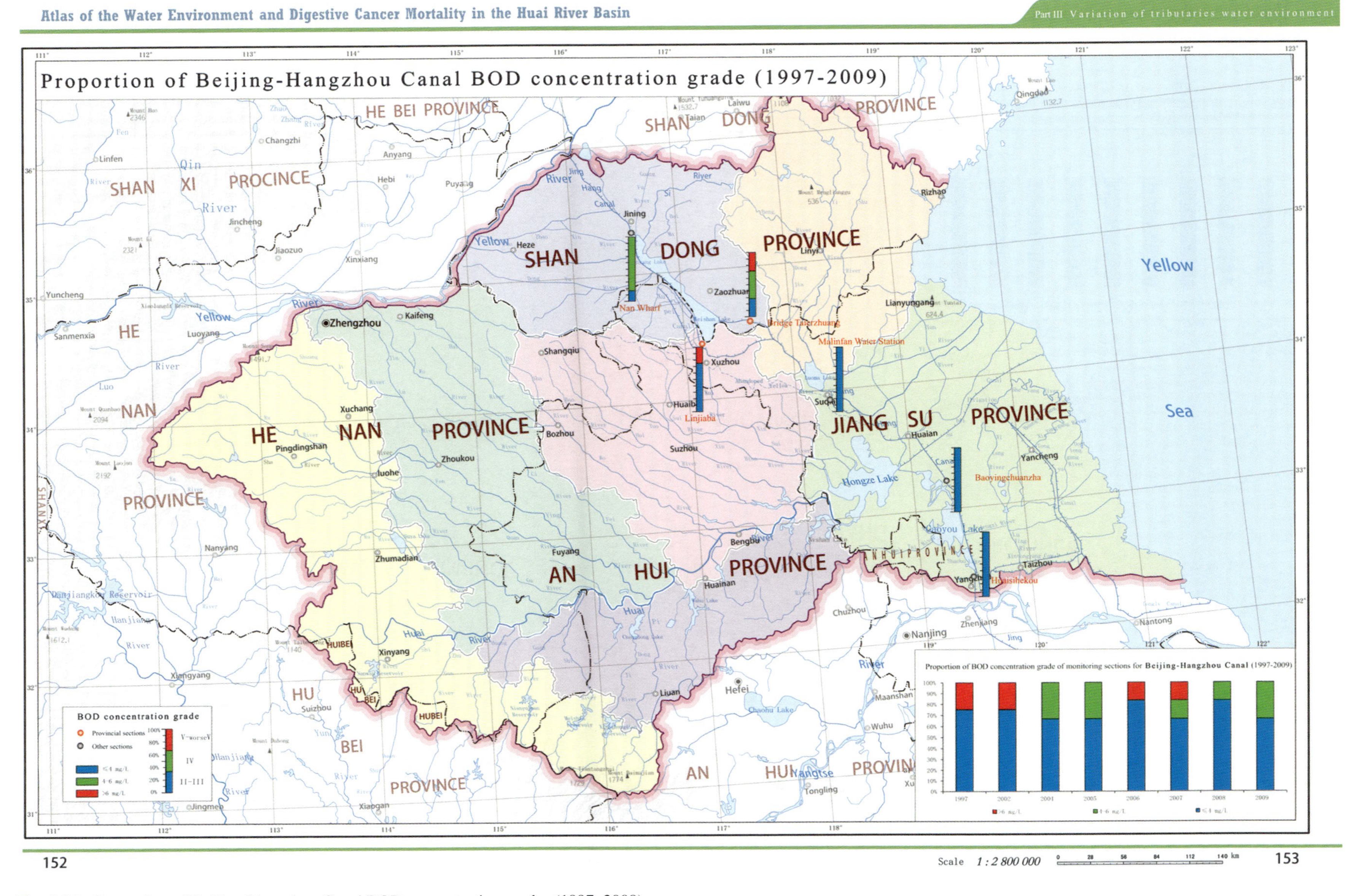

Fig. 4.36 Proportion of Beijing-Hangzhou Canal BOD concentration grades (1997–2009)

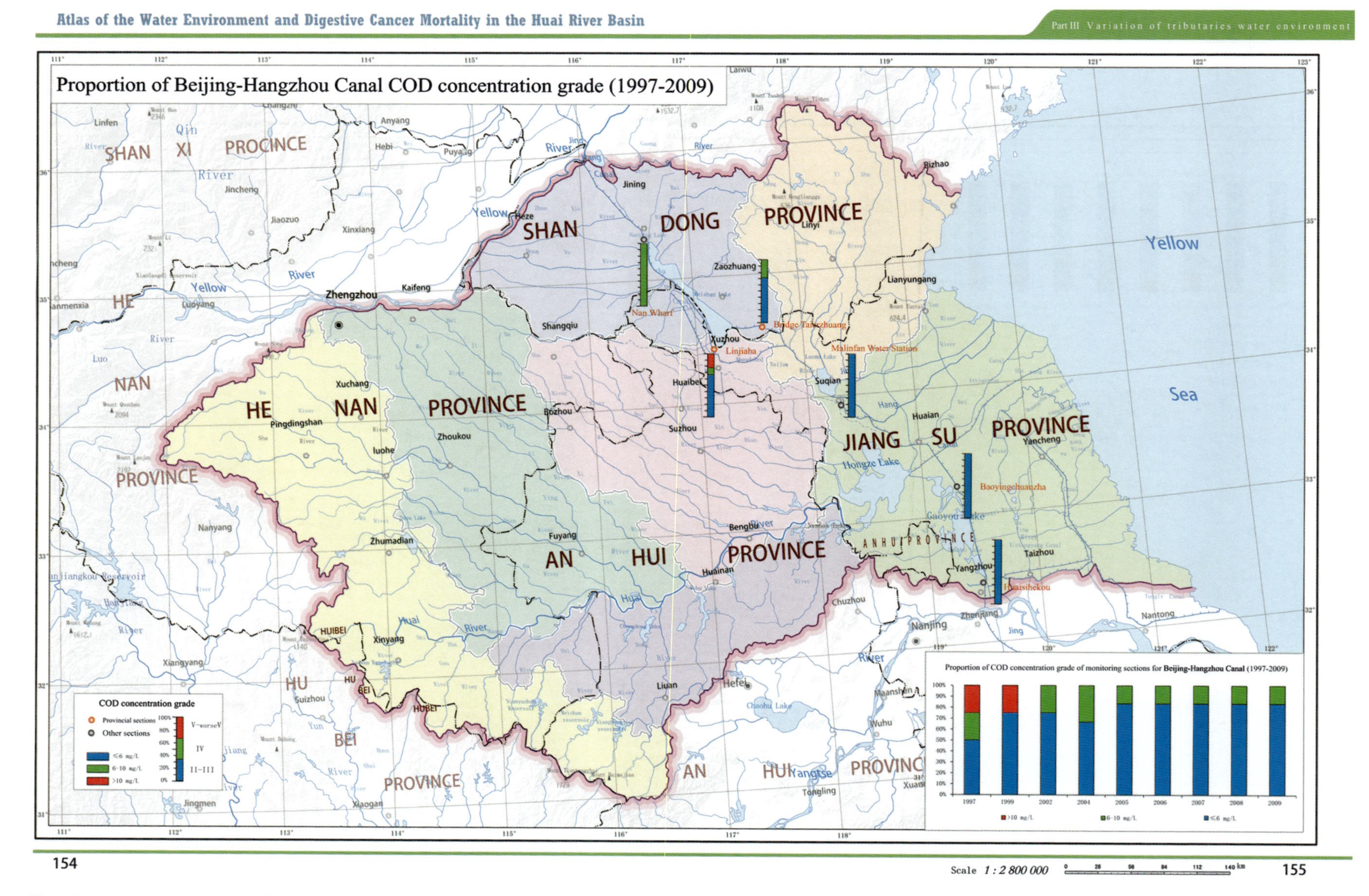

Fig. 4.37 Proportion of Beijing-Hangzhou Canal COD concentration grades (1997–2009)

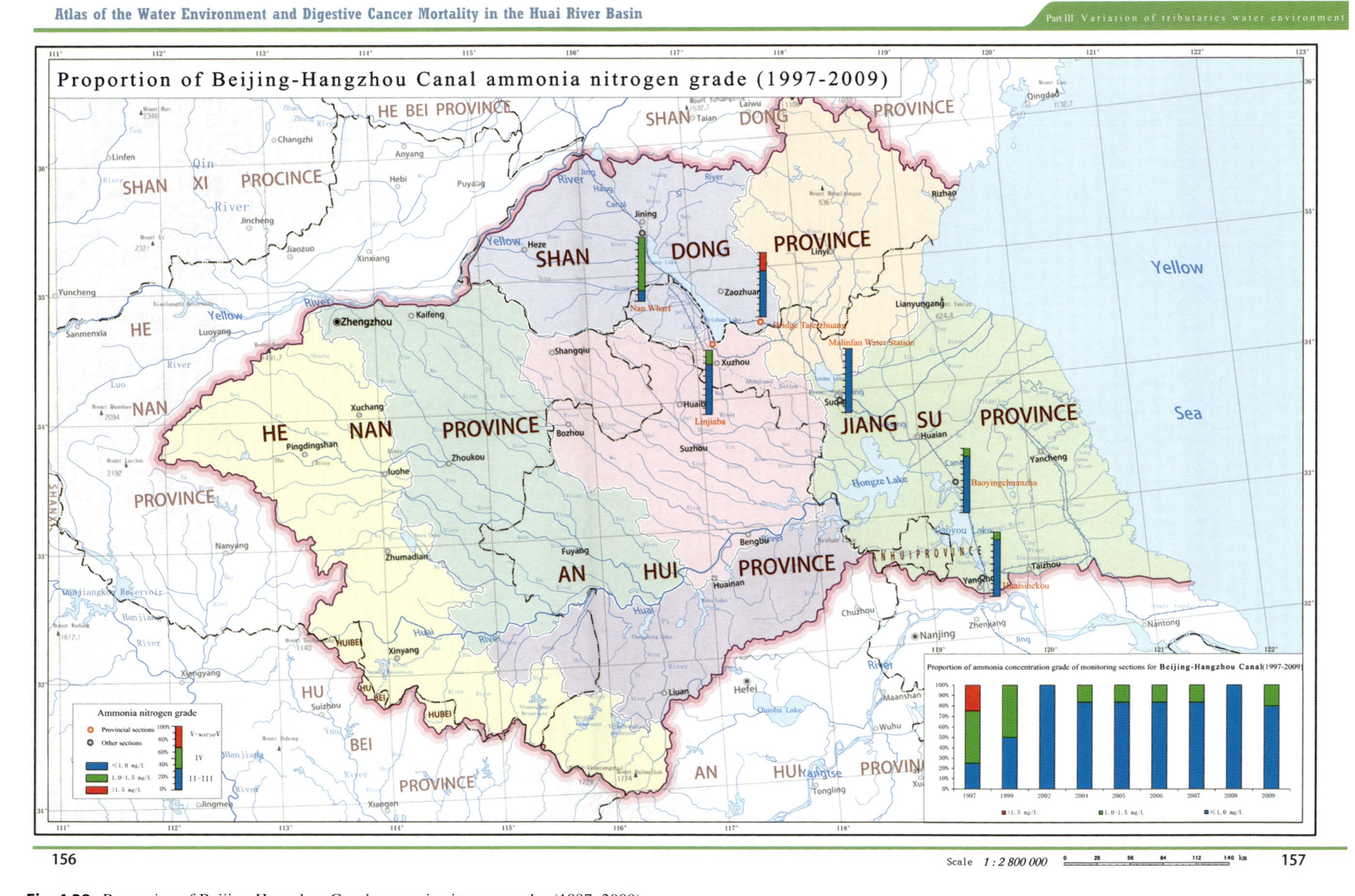

Fig. 4.38 Proportion of Beijing-Hangzhou Canal ammonia nitrogen grades (1997–2009)

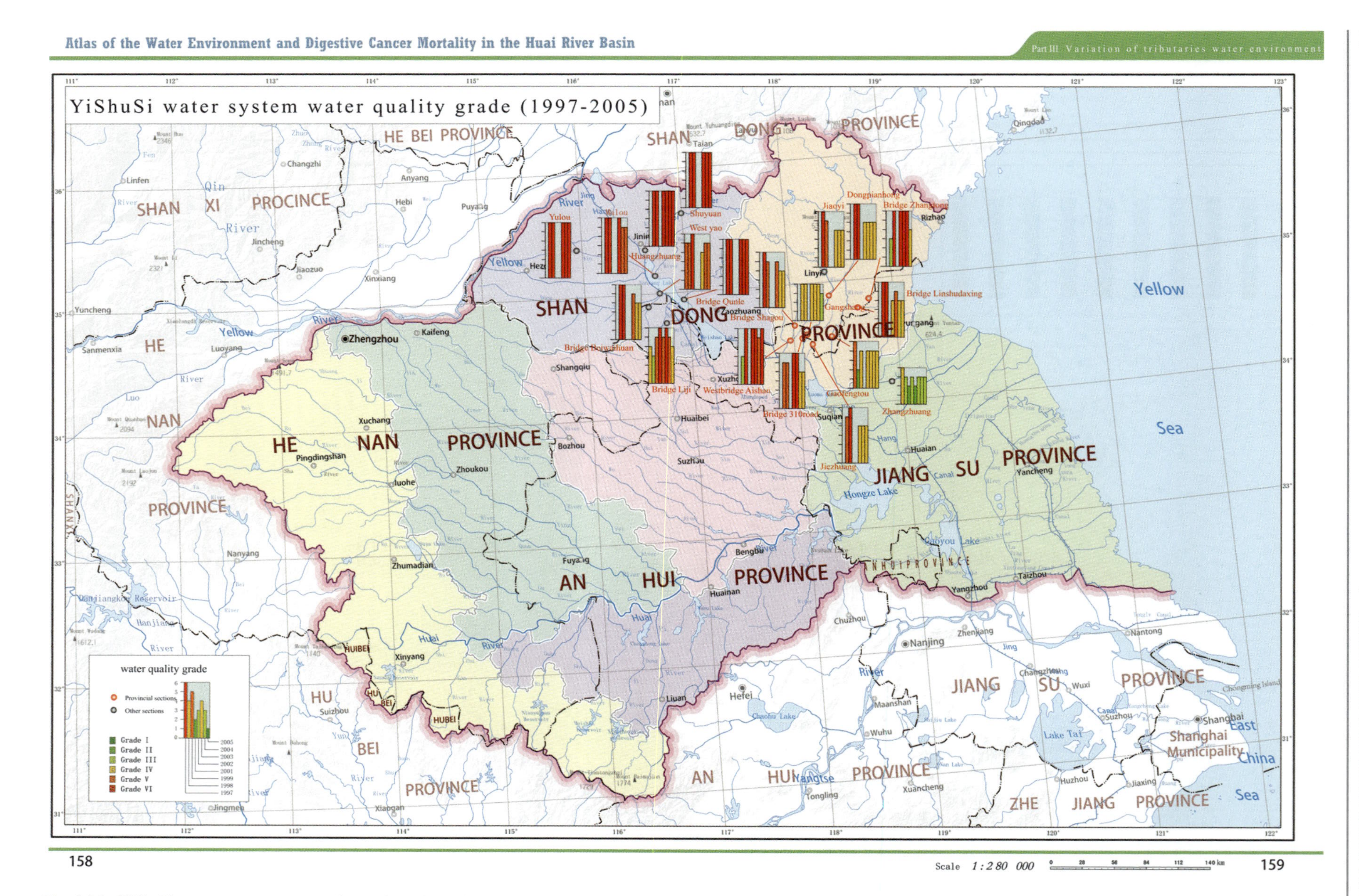

Fig. 4.39 YiShuSi water system water quality grades (1997–2005)

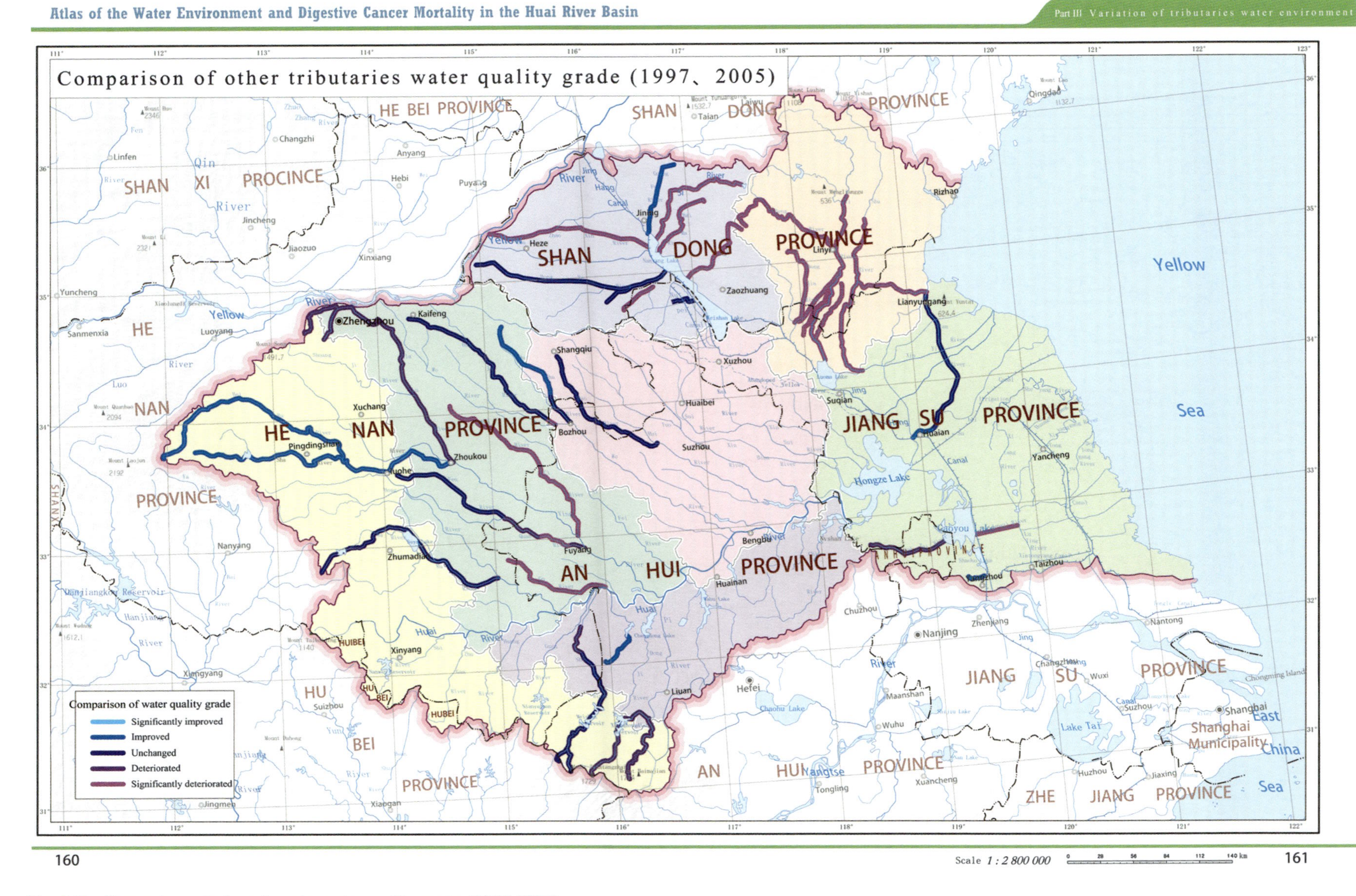

Fig. 4.40 Comparison of other tributaries water quality grades (1997–2005)

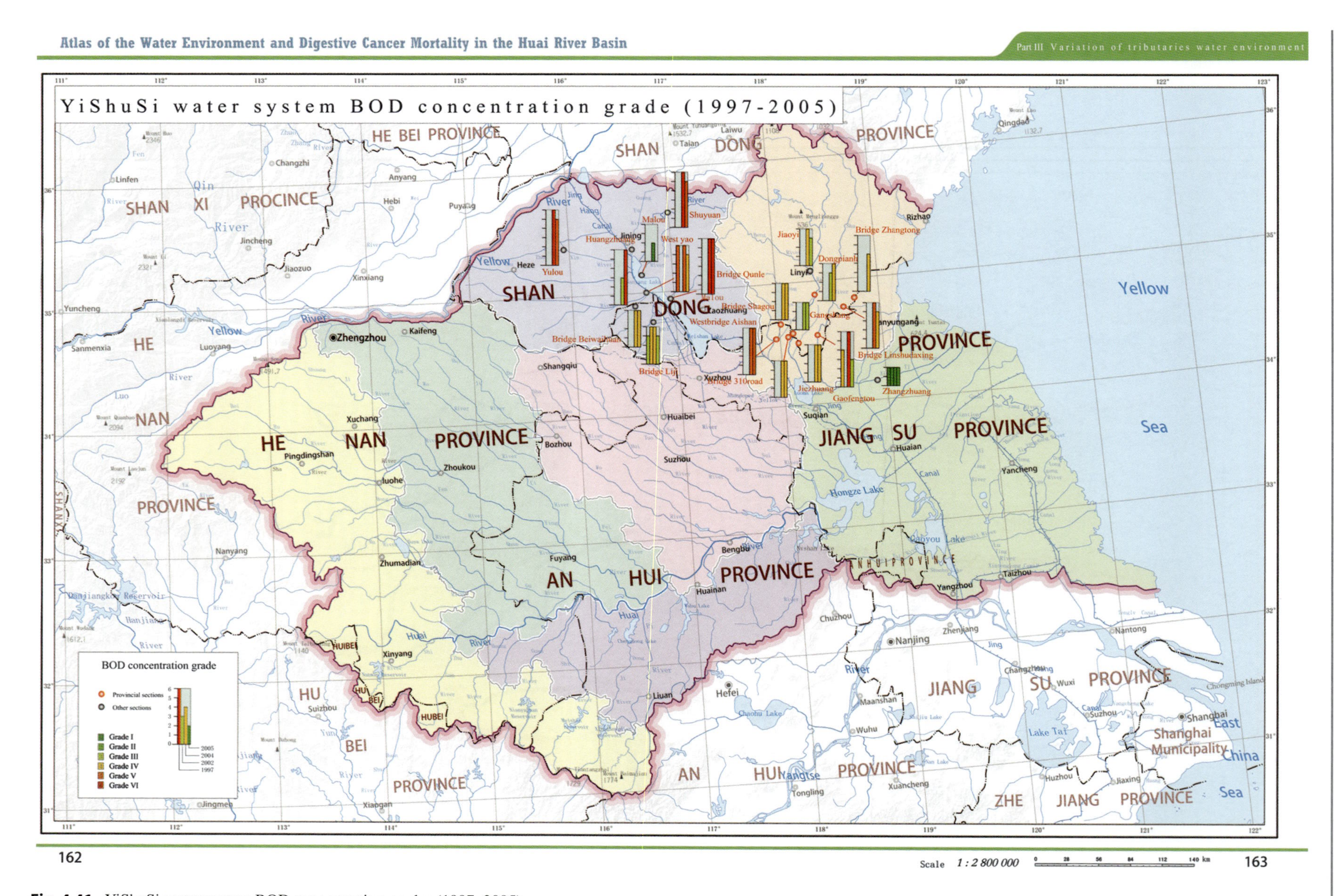

Fig. 4.41 YiShuSi water system BOD concentration grades (1997–2005)

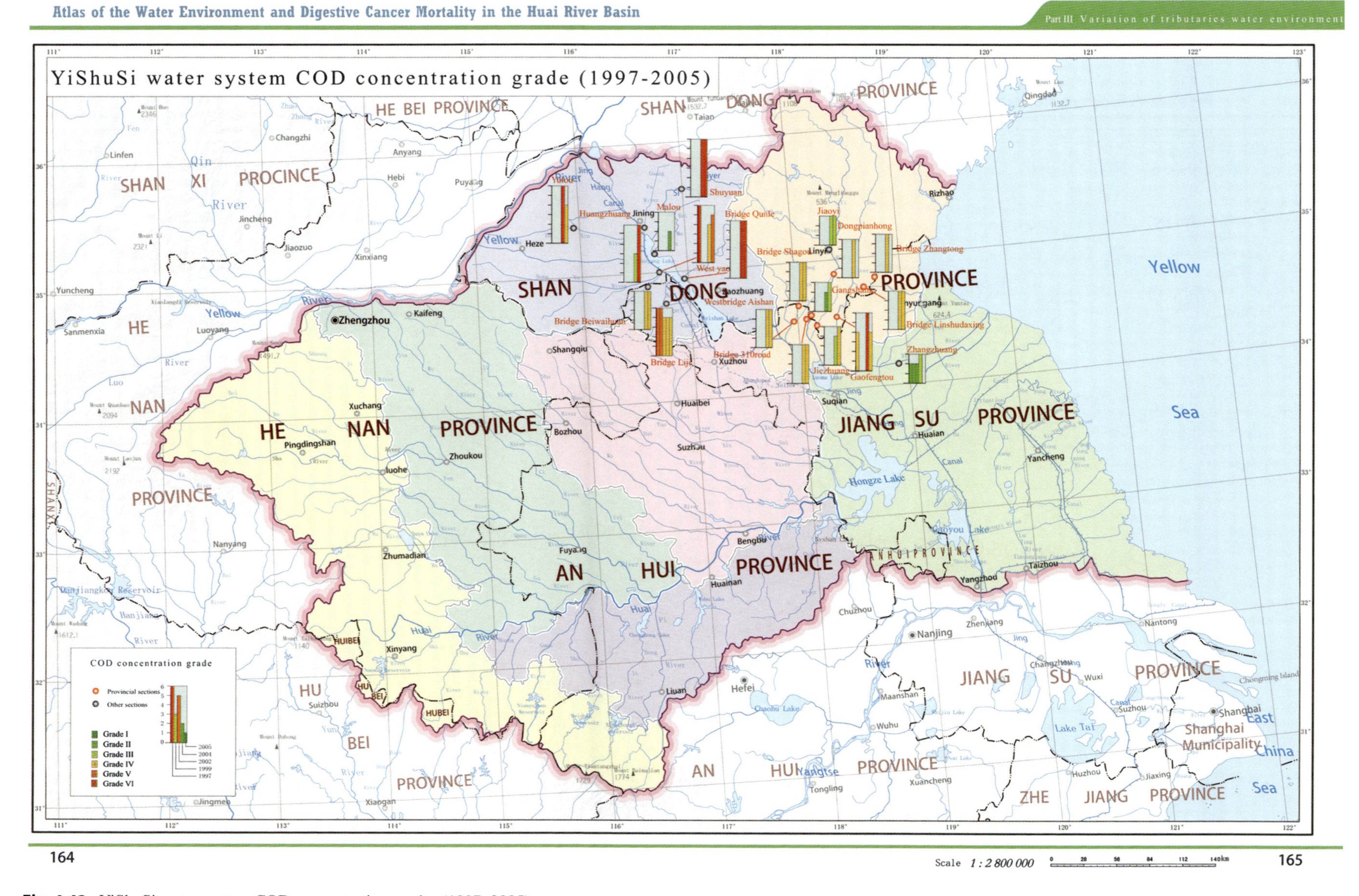

Fig. 4.42 YiShuSi water system COD concentration grades (1997–2005)

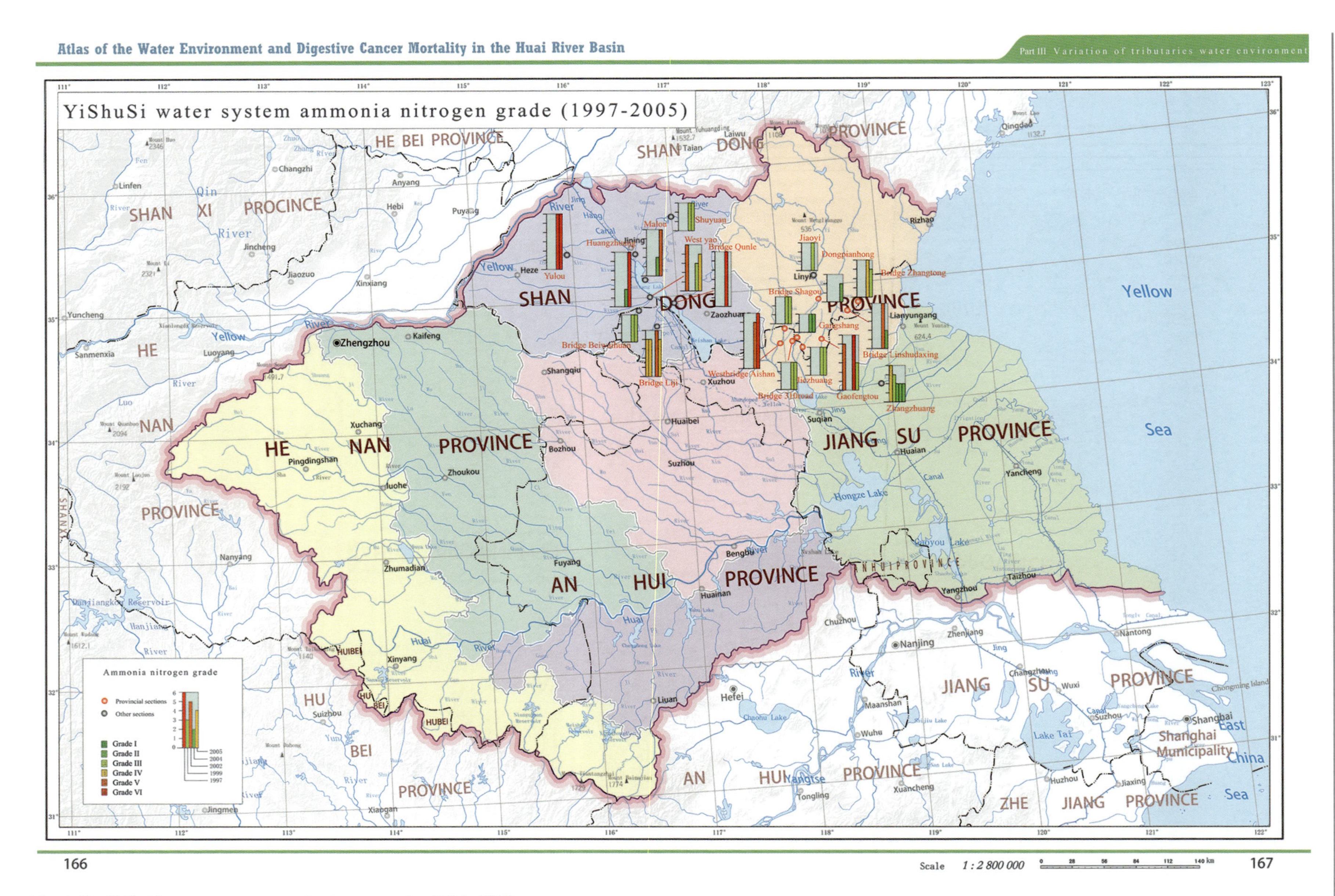

Fig. 4.43 YiShuSi water system ammonia nitrogen grades (1997–2005)

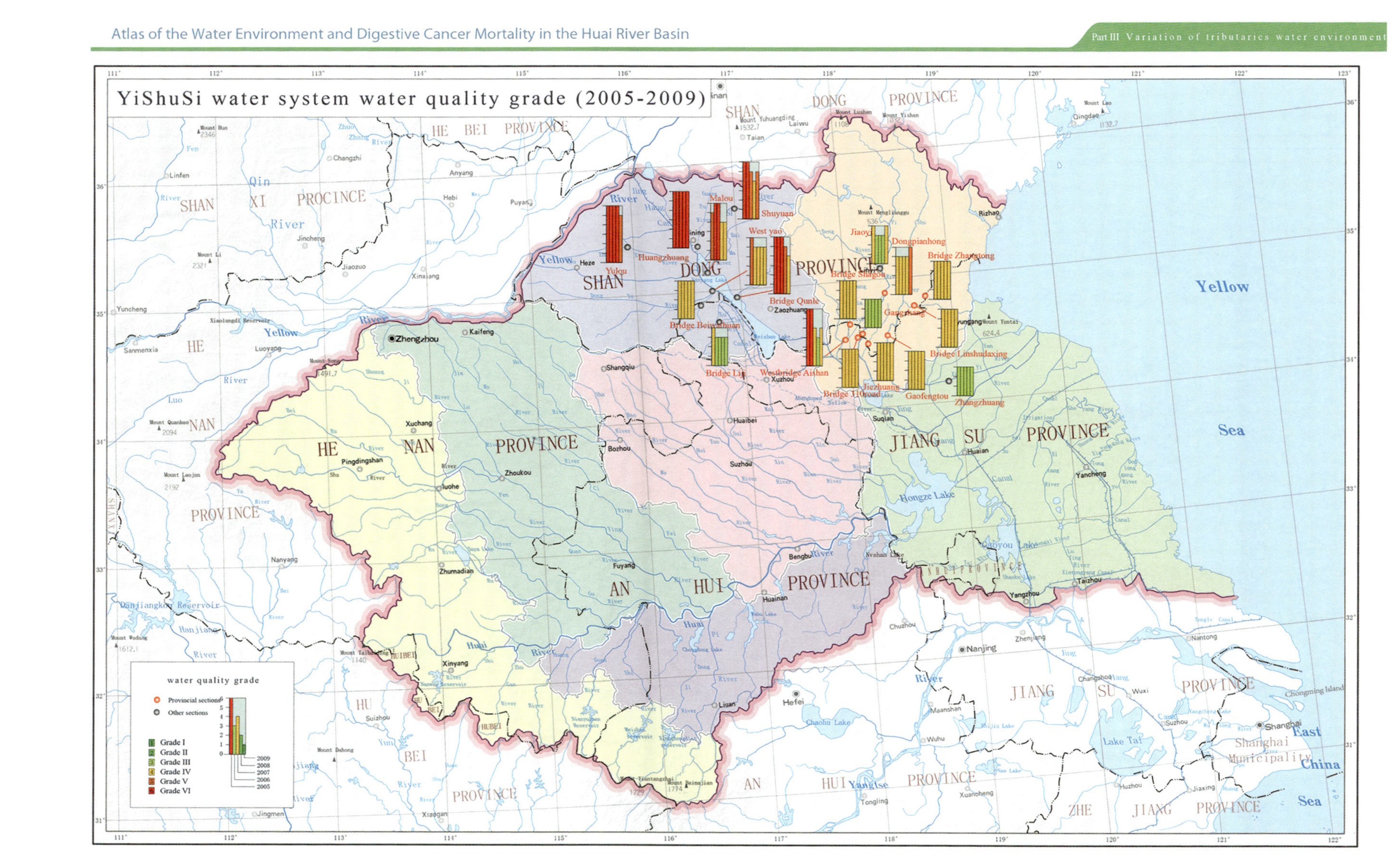

Fig. 4.44 YiShuSi water system water quality grades (2005–2009)

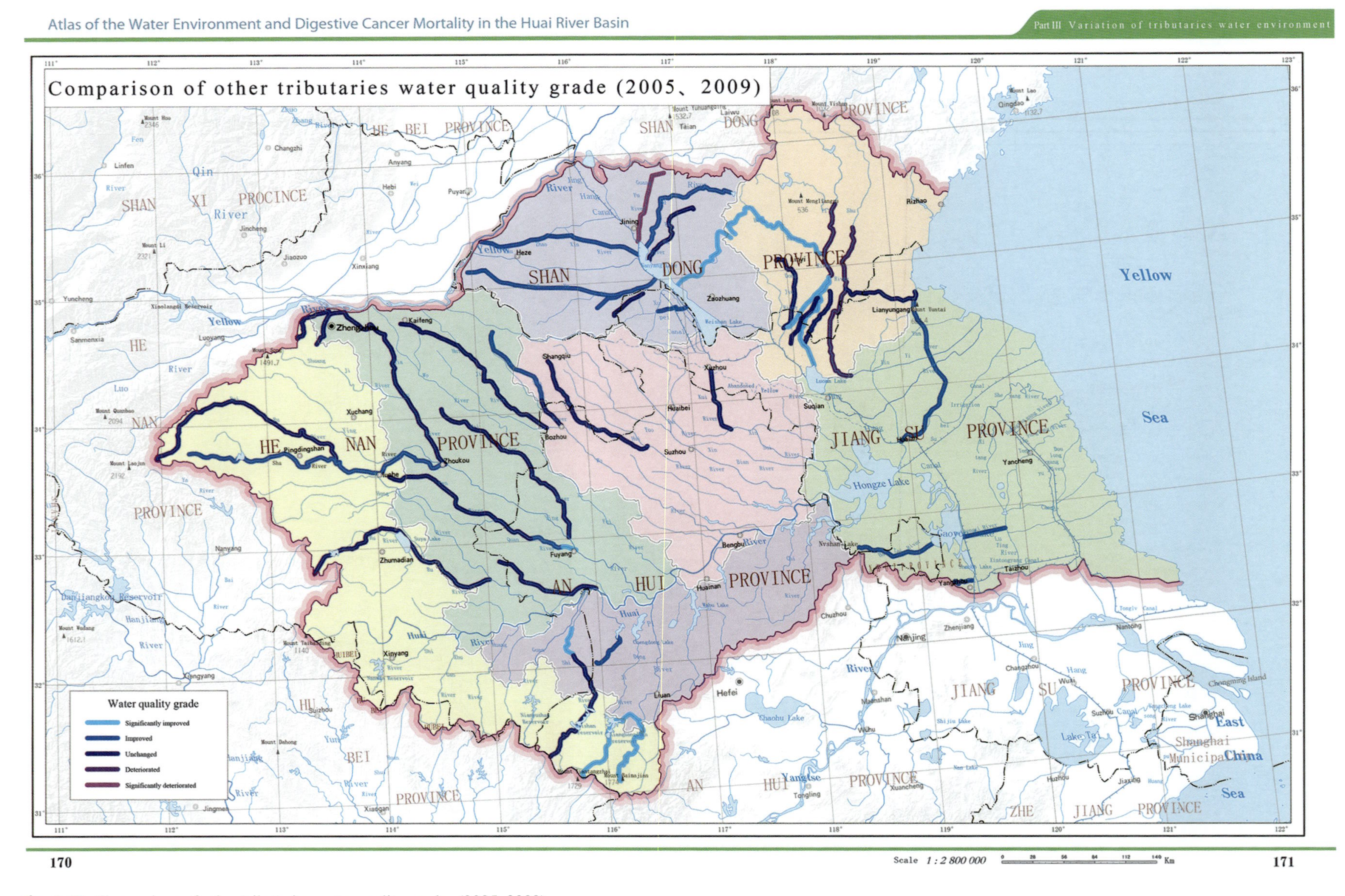

Fig. 4.45 Comparison of other tributaries water quality grades (2005–2009)

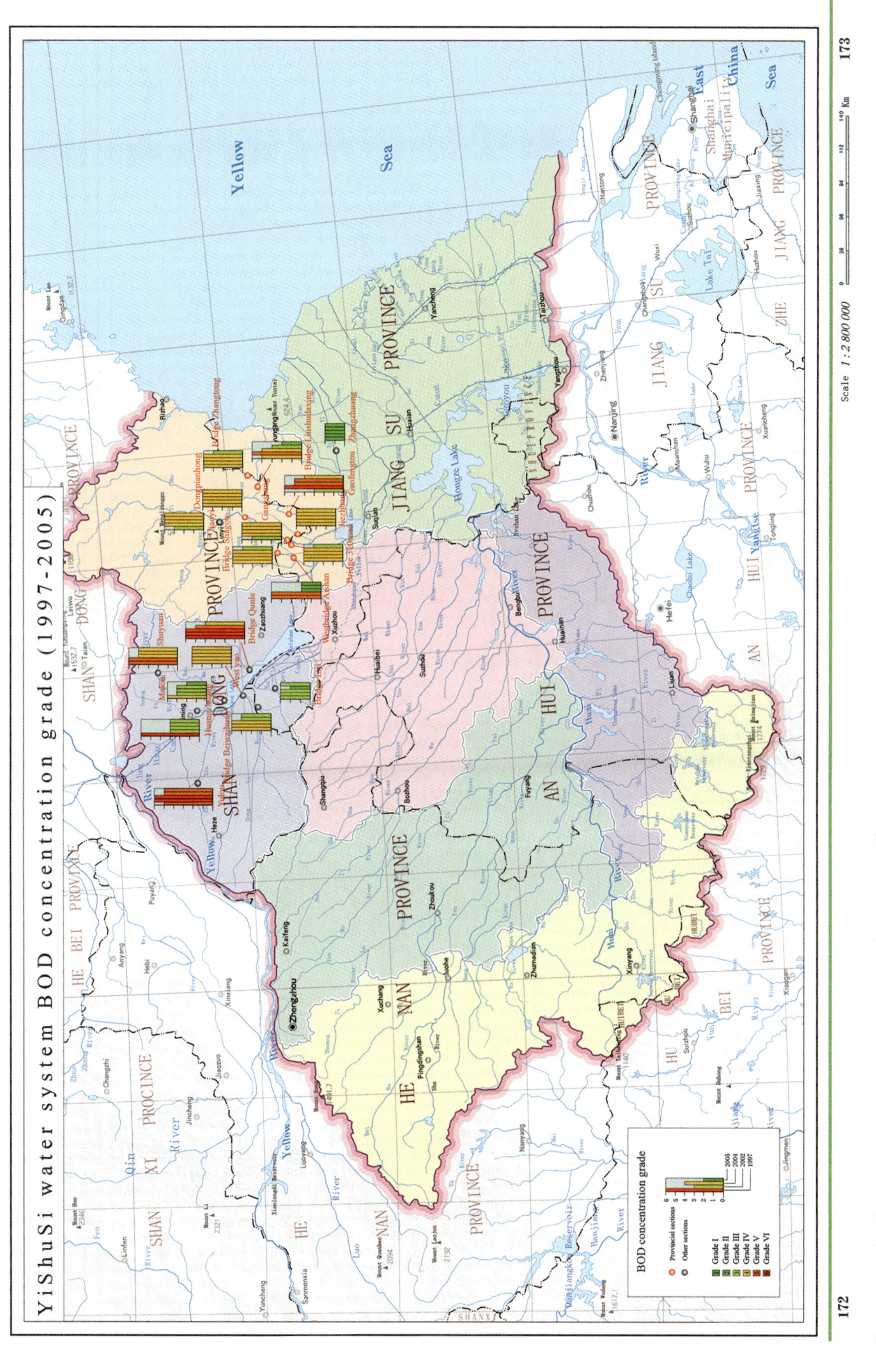

Fig. 4.46 YiShuSi water system BOD concentration grades (2005–2009)

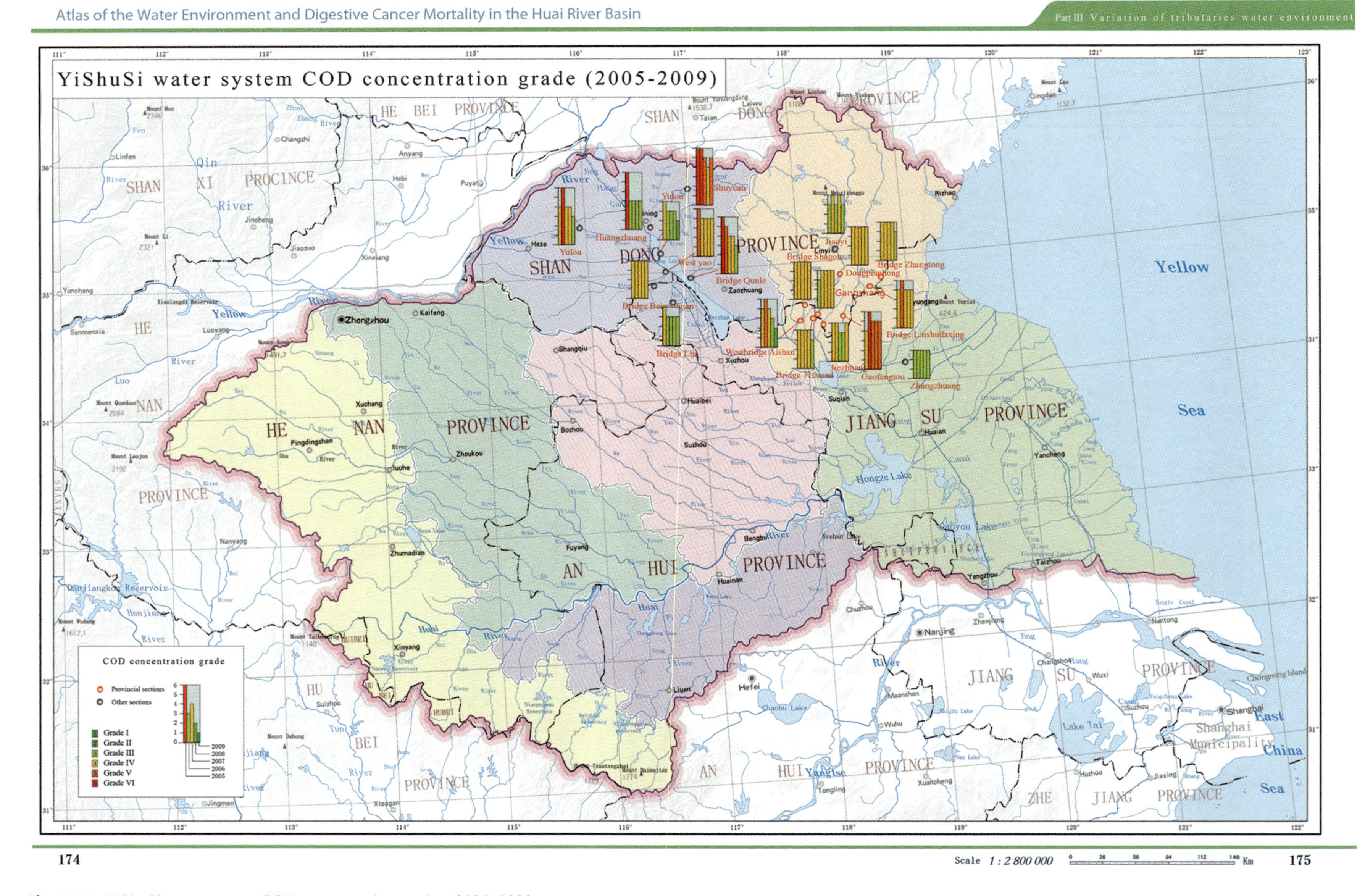

Fig. 4.47 YiShuSi water system COD concentration grades (2005–2009)

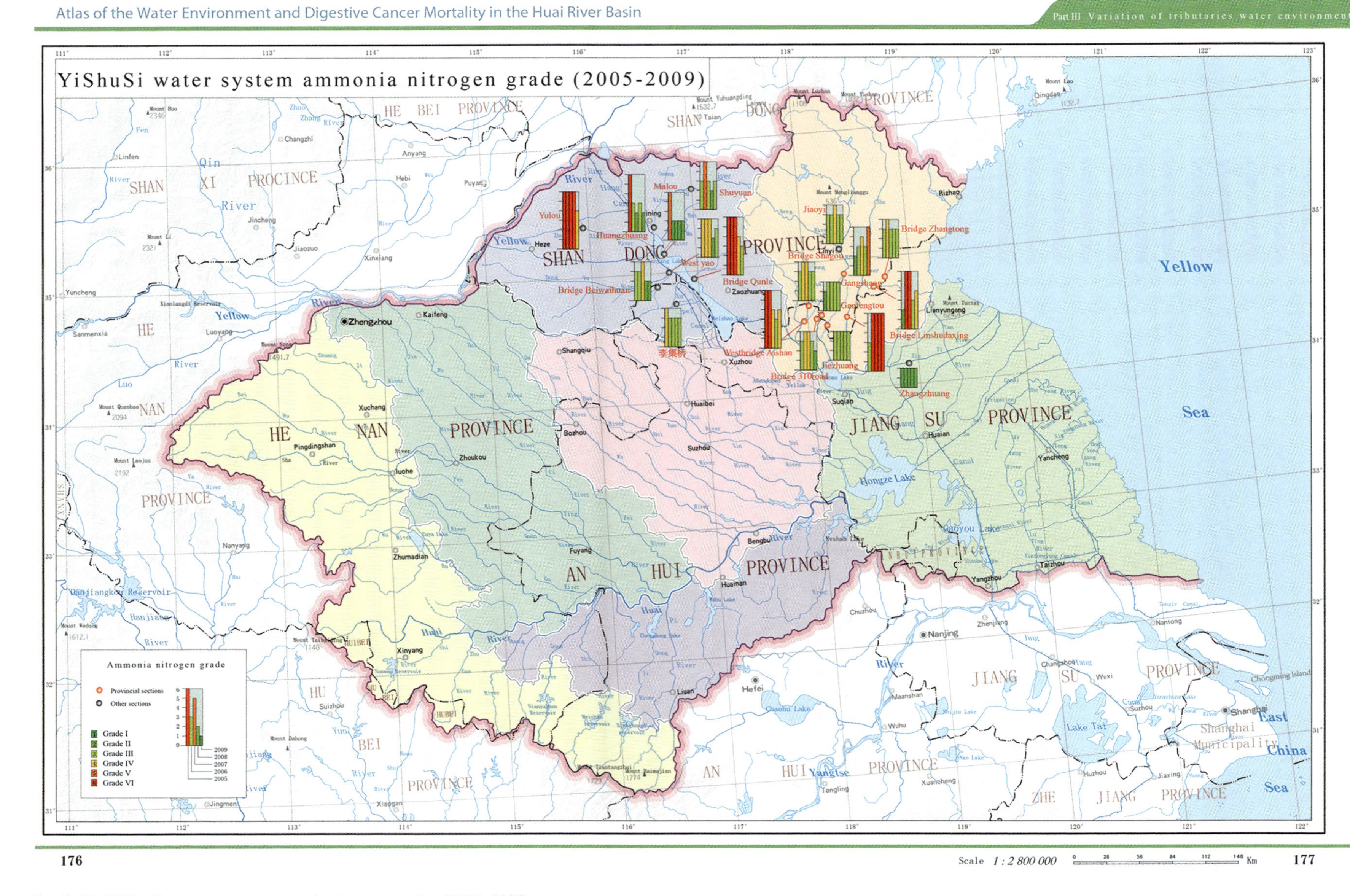

Fig. 4.48 YiShuSi water system ammonia nitrogen grades (2005–2009)

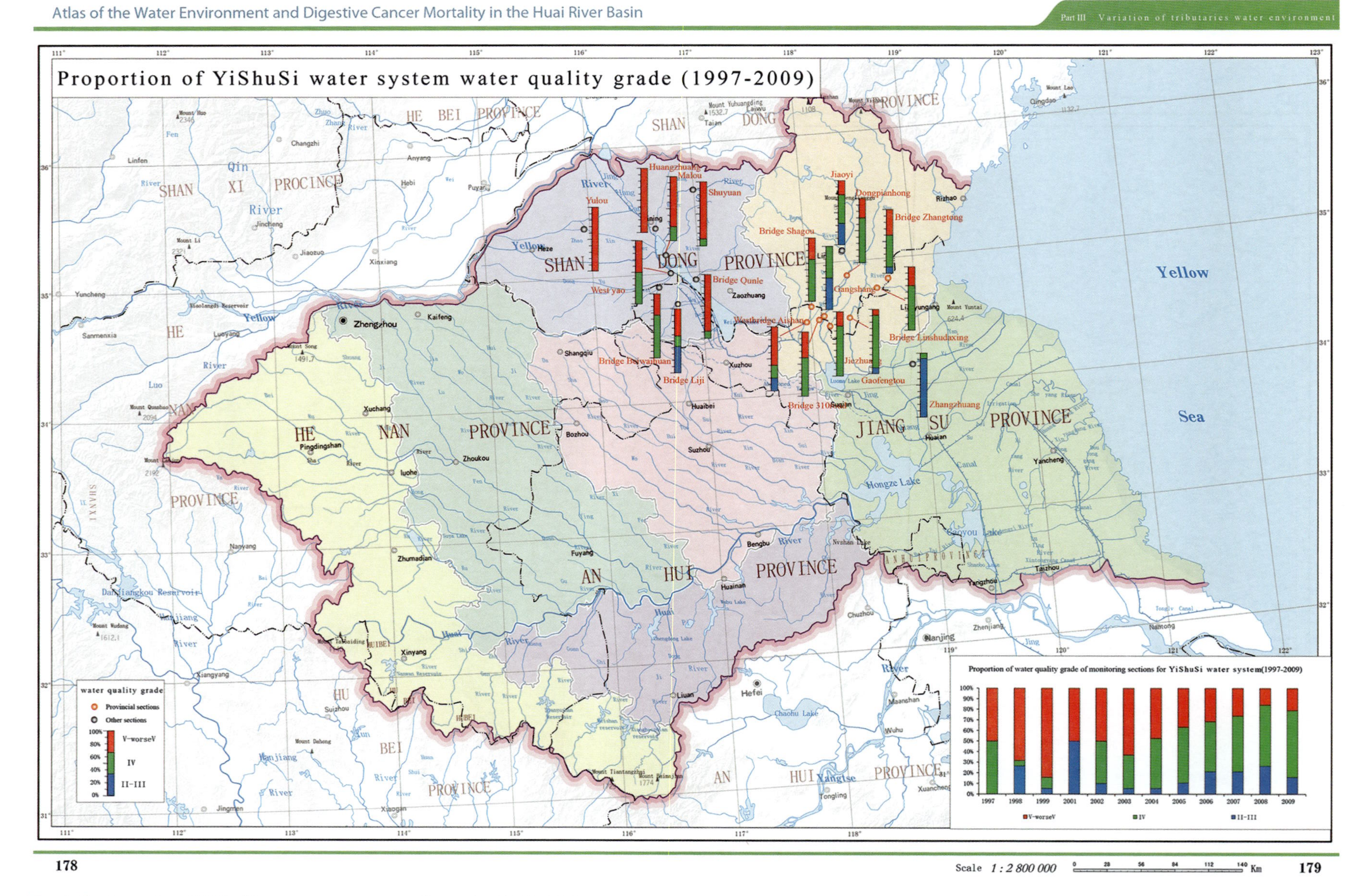

Fig. 4.49 Proportion of YiShuSi water system water quality grades (1997–2009)

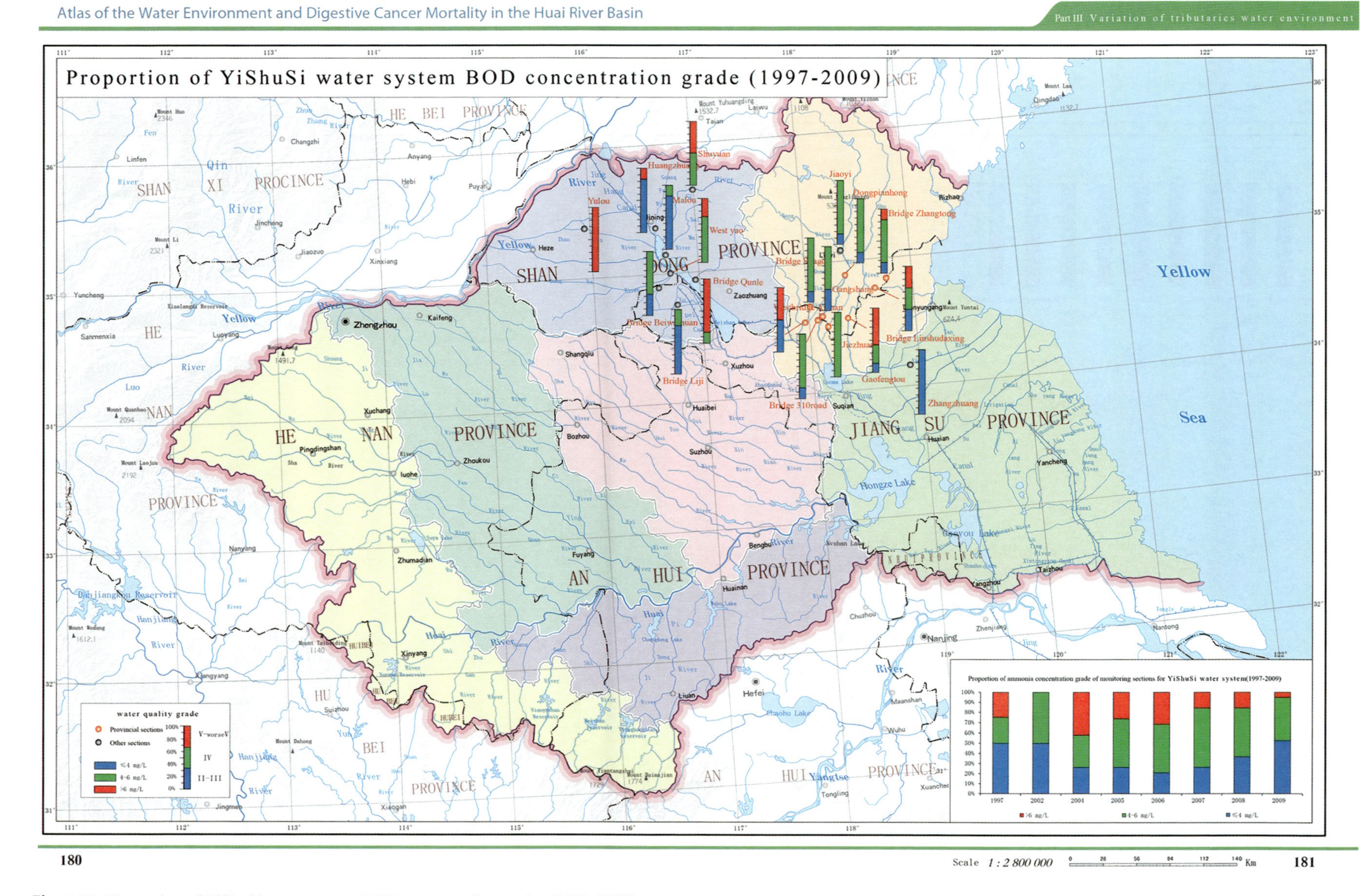

Fig. 4.50 Proportion of YiShuSi water system BOD concentration grades (1997–2009)

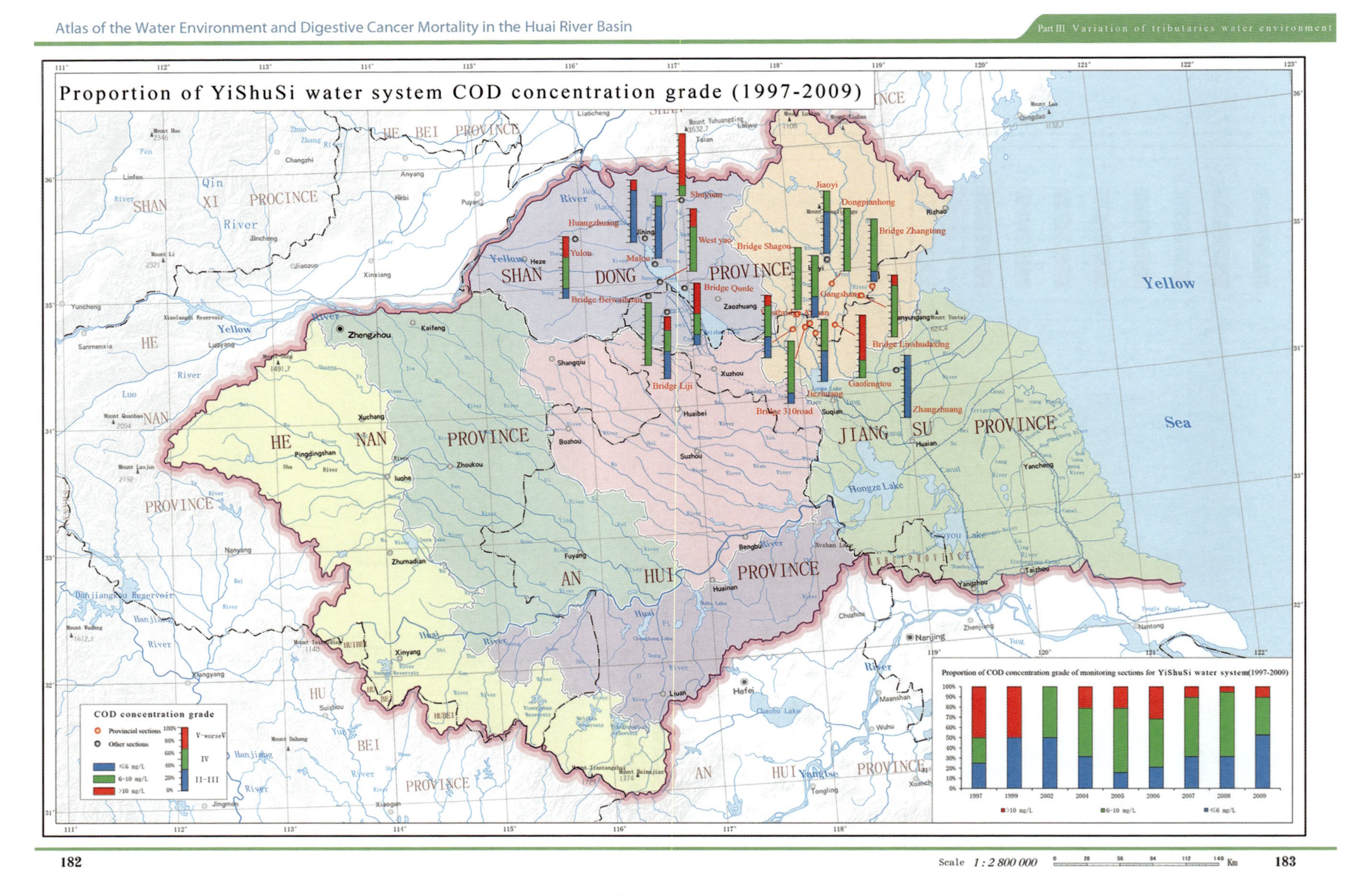

Fig. 4.51 Proportion of YiShuSi water system COD concentration grades (1997–2009)

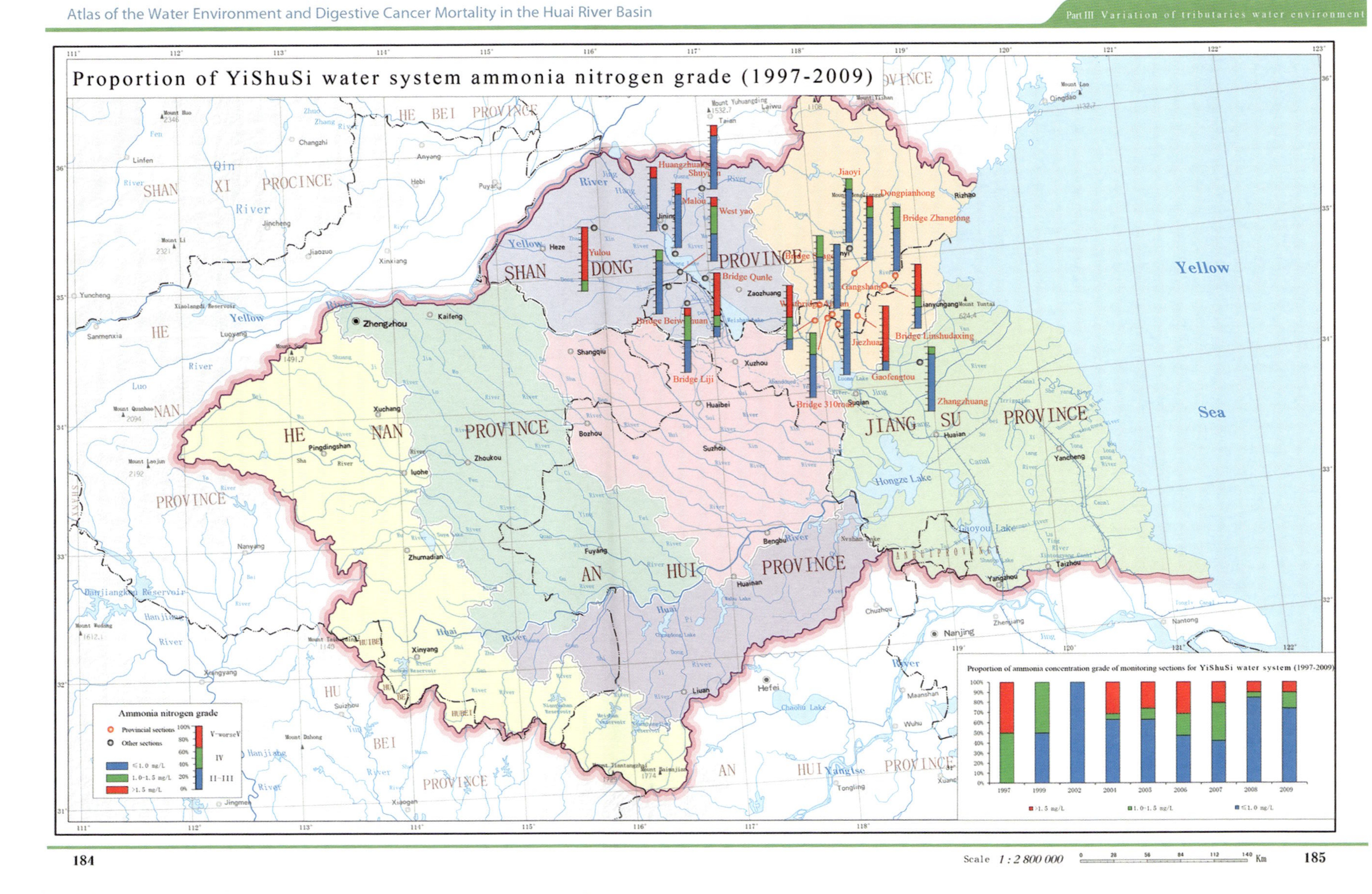

Fig. 4.52 Proportion of YiShuSi water system ammonia nitrogen grades (1997–2009)

5 Variation in the Water Environment of Lakes

G. Yang, D. Zhuang (eds.), *Atlas of the Huai River Basin Water Environment: Digestive Cancer Mortality*,
DOI 10.1007/978-94-017-8619-5_5,

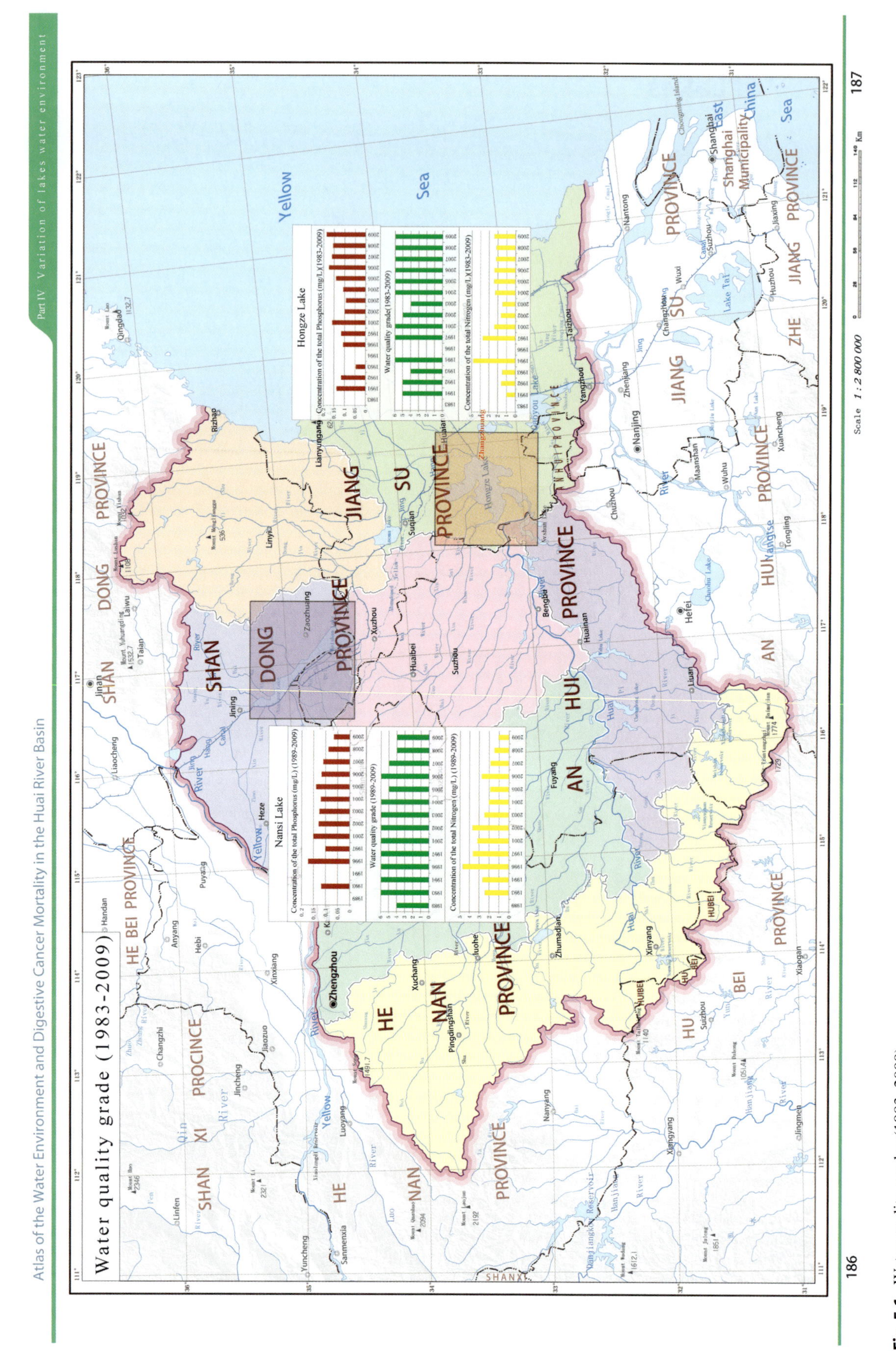

Fig. 5.1 Water quality grades (1983–2009)

Spatiotemporal Variation in the Frequency of Water Pollution

6

G. Yang, D. Zhuang (eds.), *Atlas of the Huai River Basin Water Environment: Digestive Cancer Mortality*,
DOI 10.1007/978-94-017-8619-5_6, © Springer Science+Business Media Dordrecht 2014

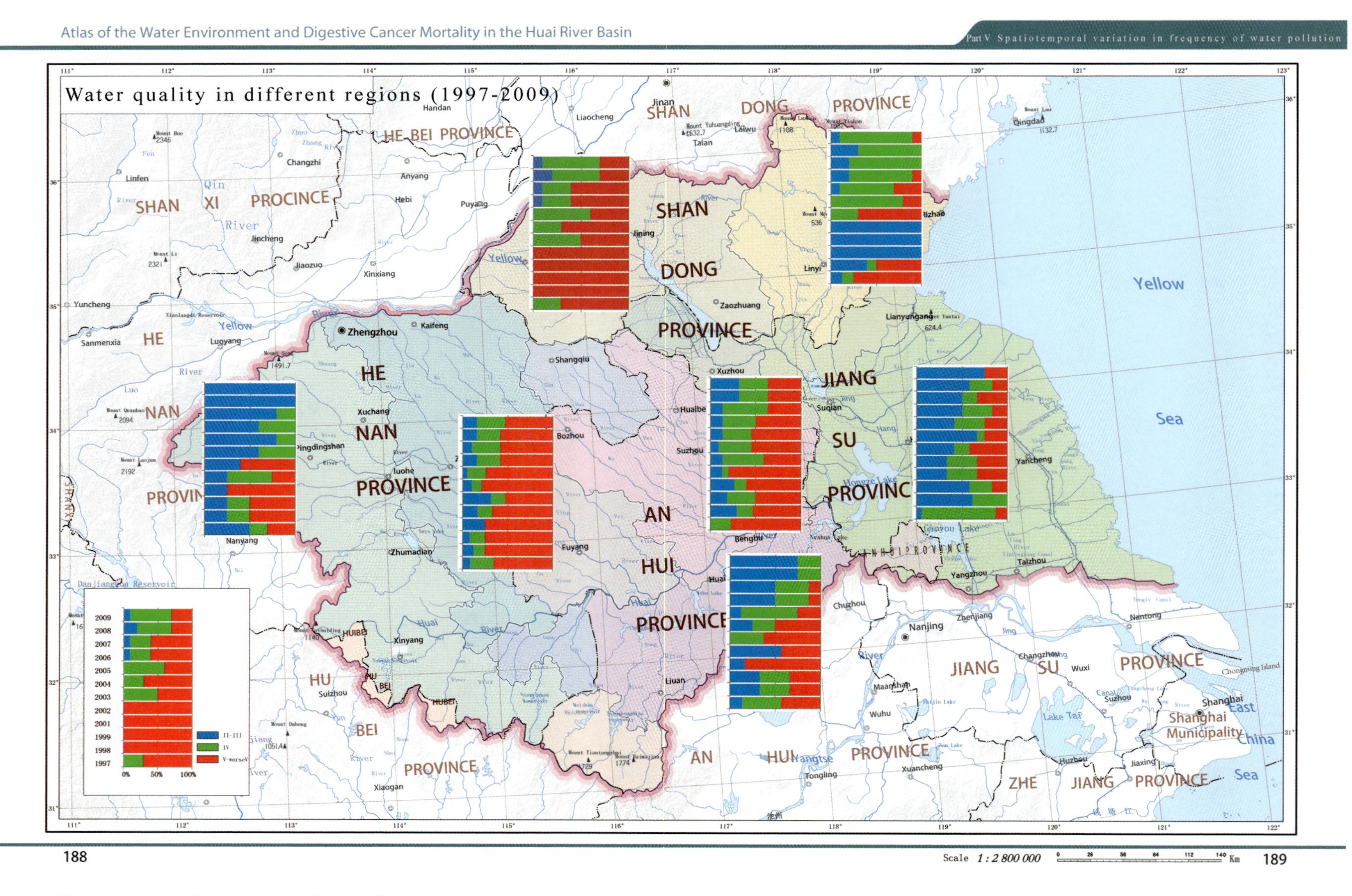

Fig. 6.1 Water quality in different regions (1997–2009)

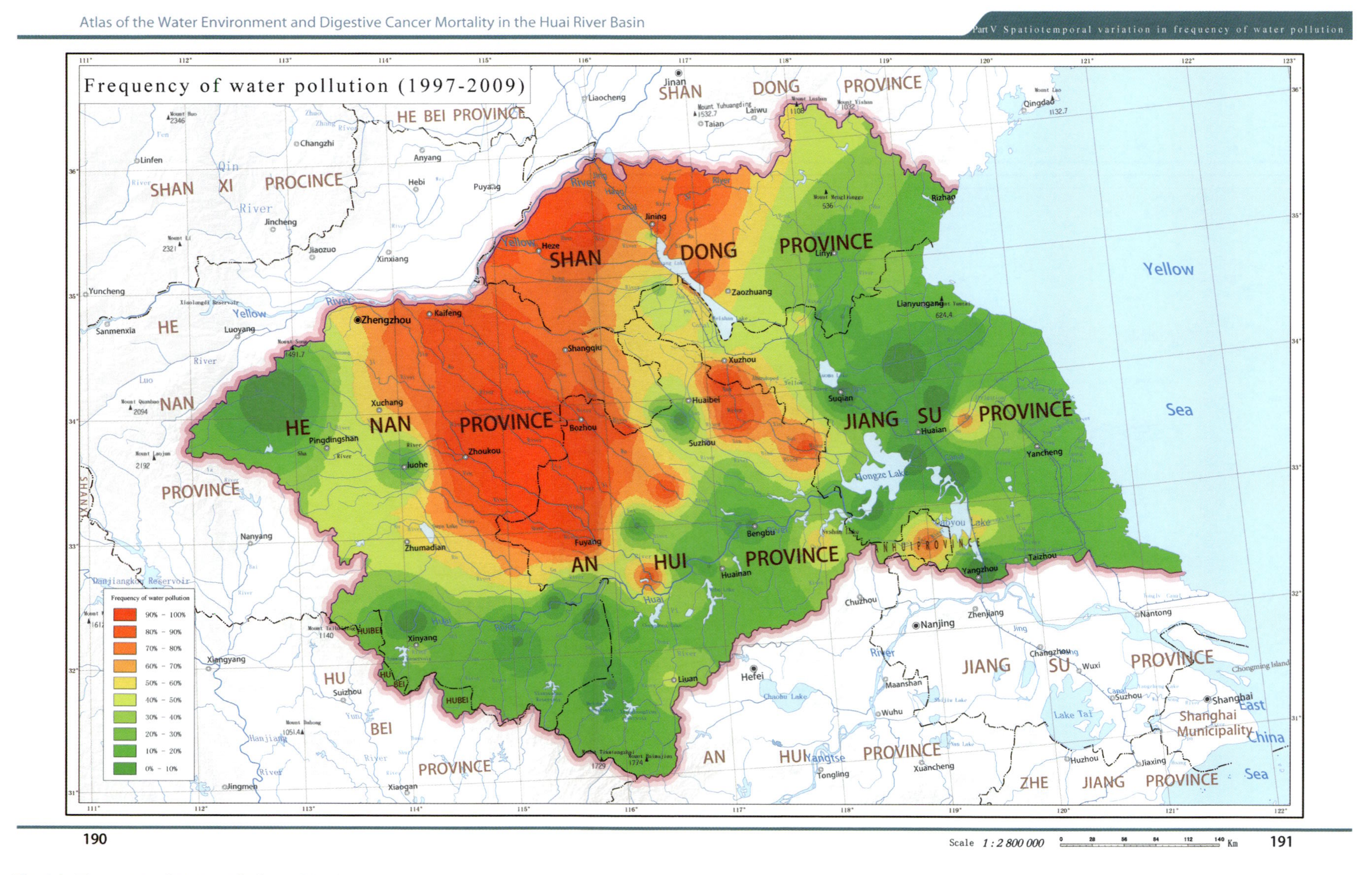

Fig. 6.2 Frequency of water pollution (1997–2009)

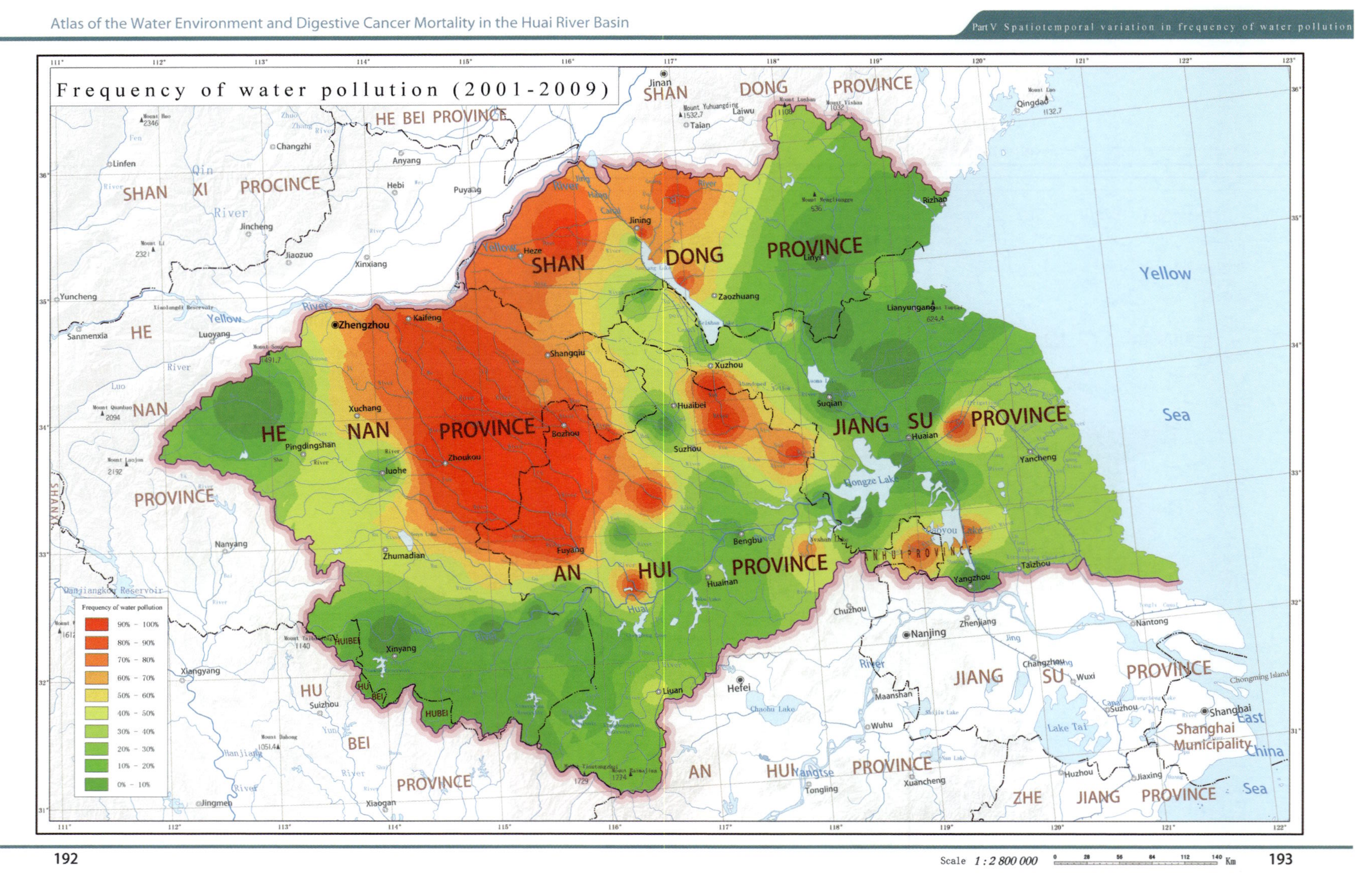

Fig. 6.3 Frequency of water pollution (2001–2009)

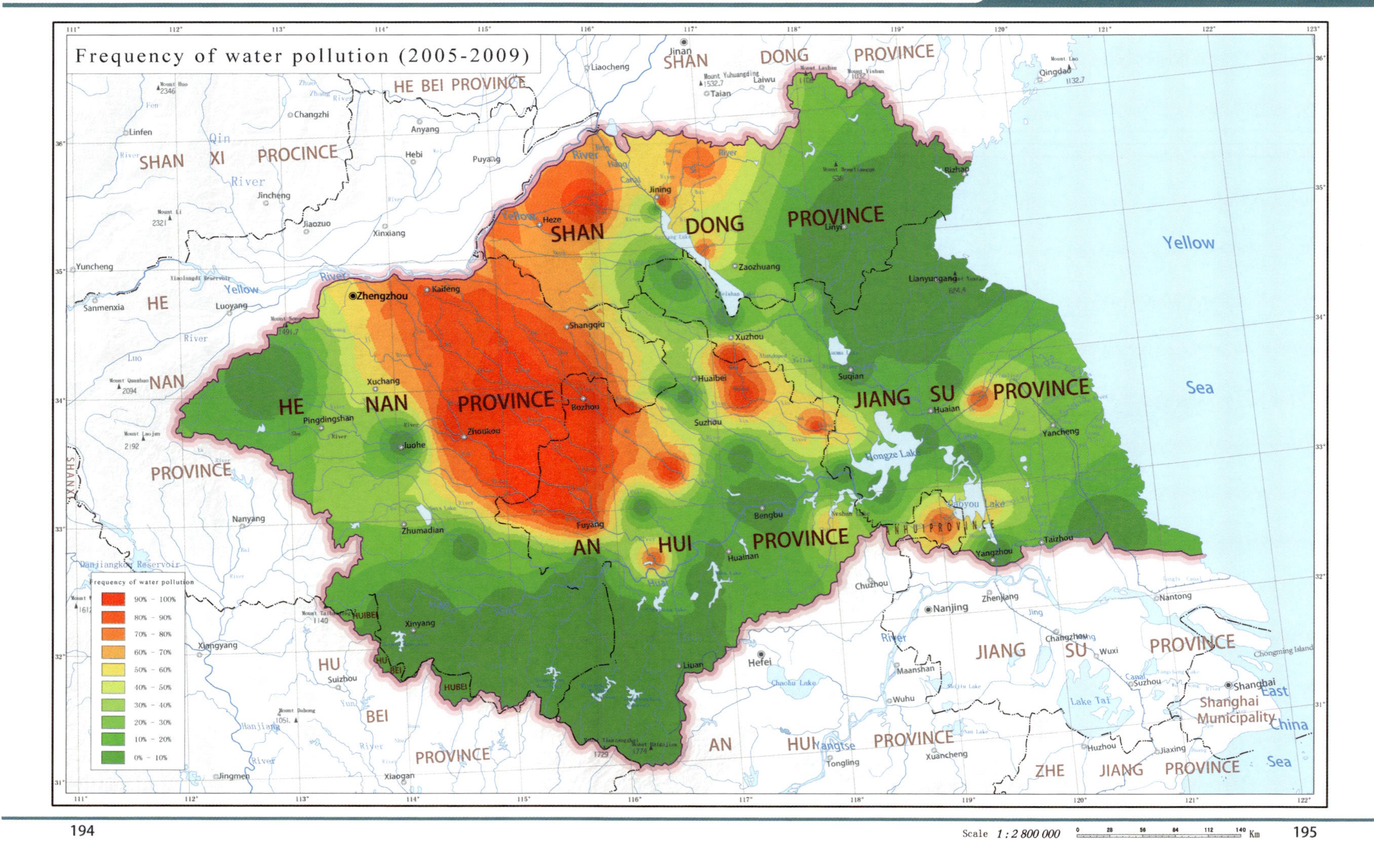

Fig. 6.4 Frequency of water pollution (2005–2009)

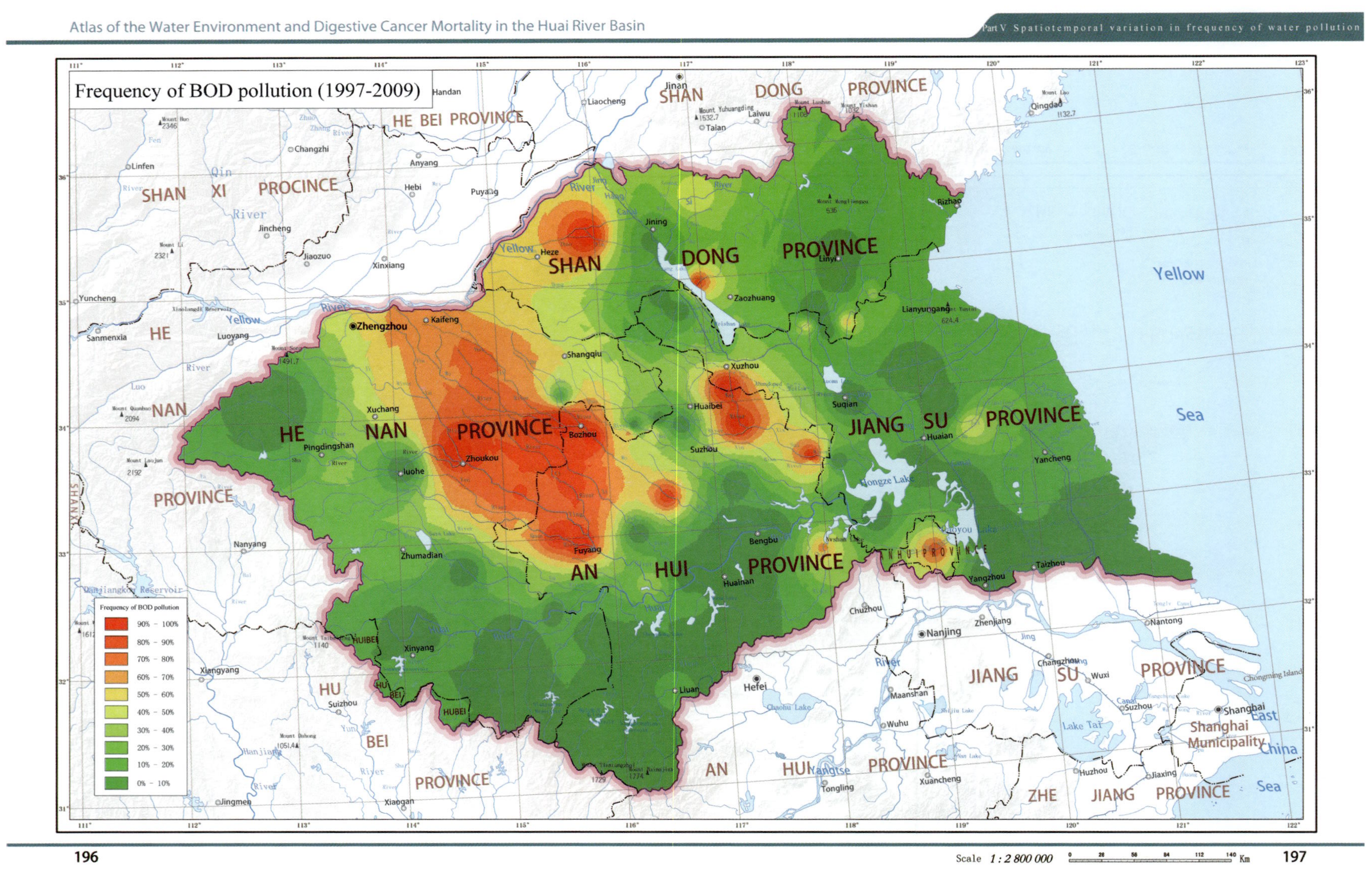

Fig. 6.5 Frequency of BOD pollution (1997–2009)

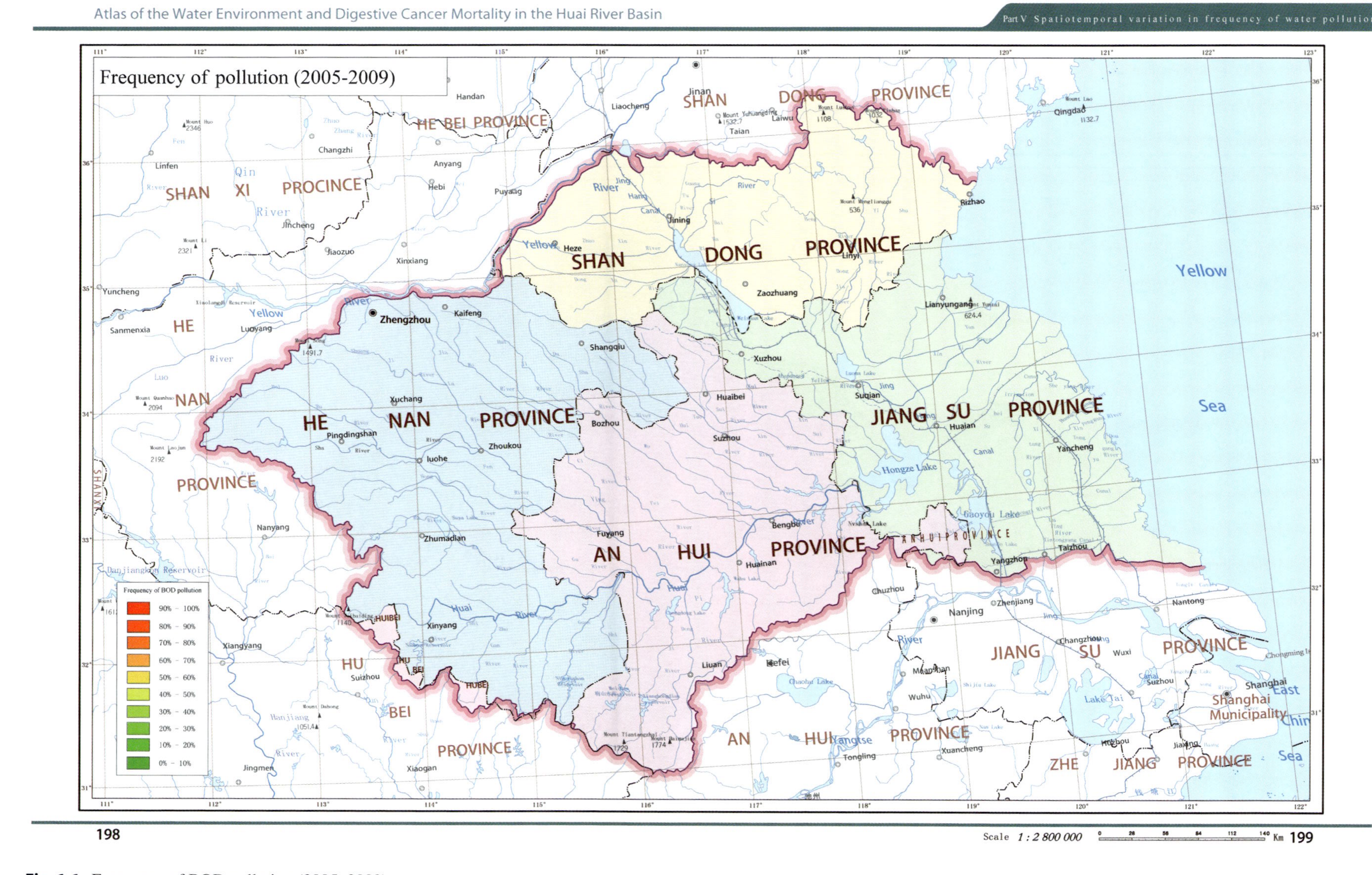

Fig. 6.6 Frequency of BOD pollution (2005–2009)

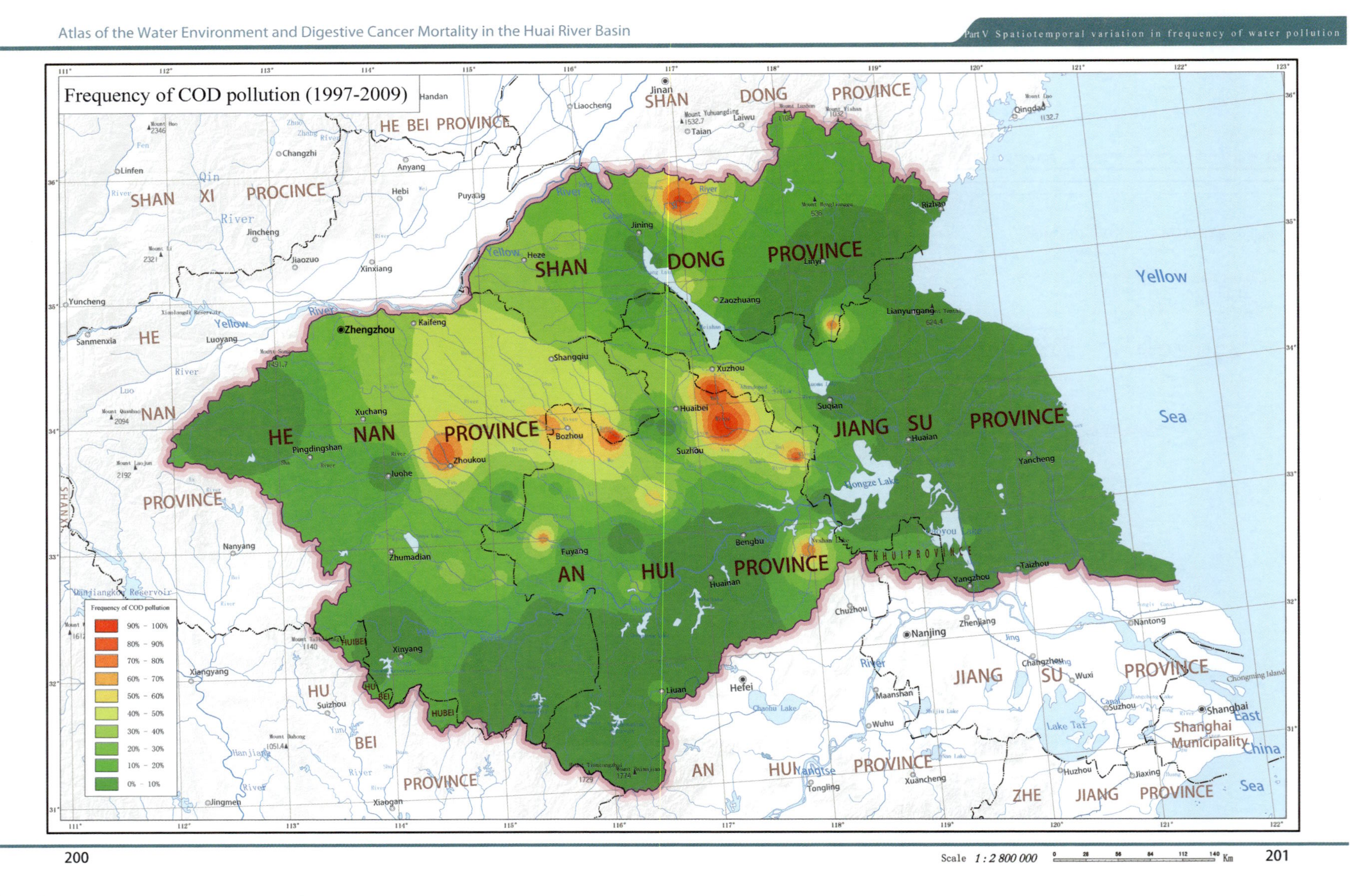

Fig. 6.7 Frequency of COD pollution (1997–2009)

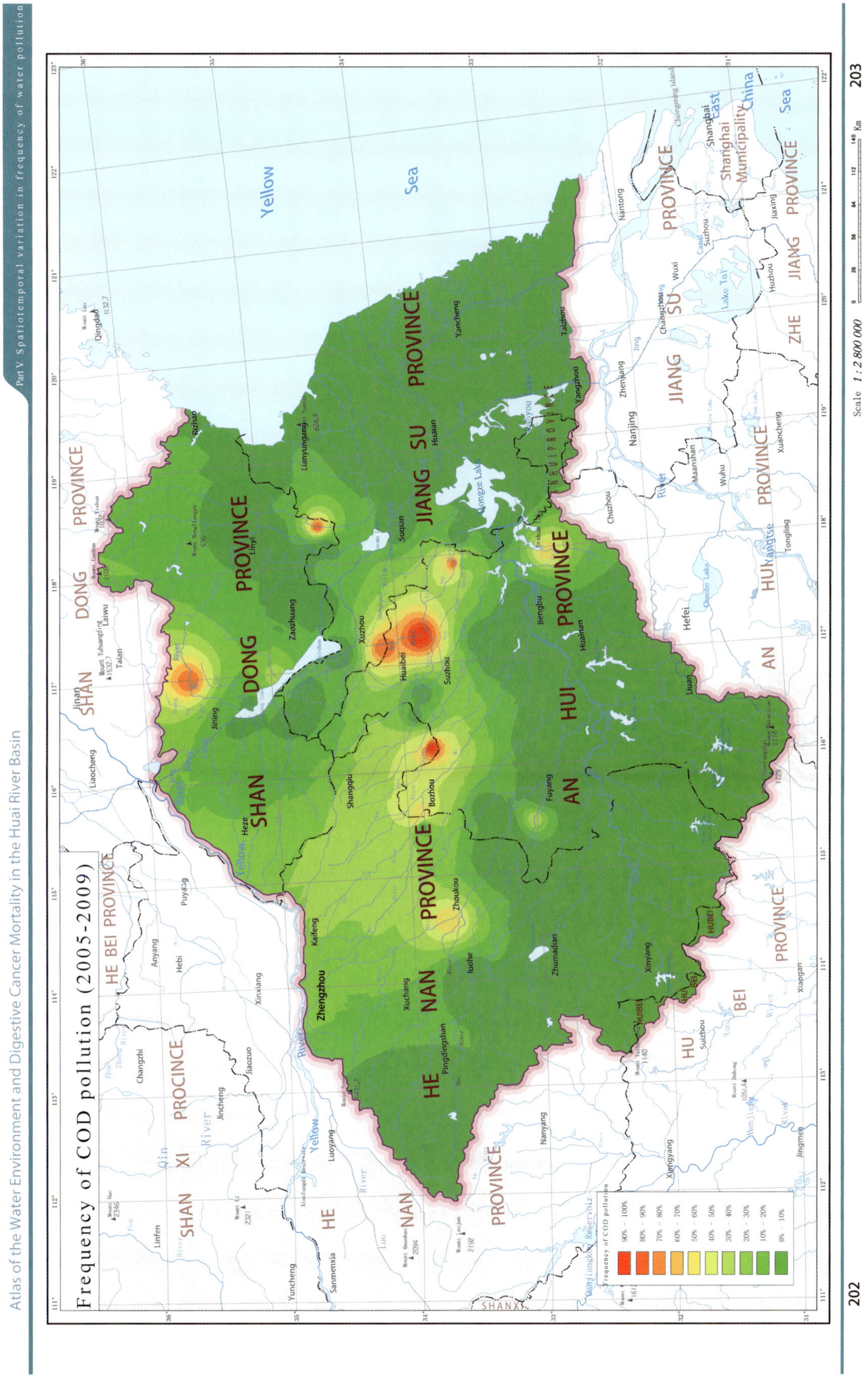

Frequency of COD pollution (2005-2009)

Frequency of COD pollution

90% - 100%
80% - 90%
70% - 80%
60% - 70%
50% - 60%
40% - 50%
30% - 40%
20% - 30%
10% - 20%
0% - 10%

Scale 1 : 2 800 000

202 203

Fig. 6.8 Frequency of COD pollution (2005–2009)

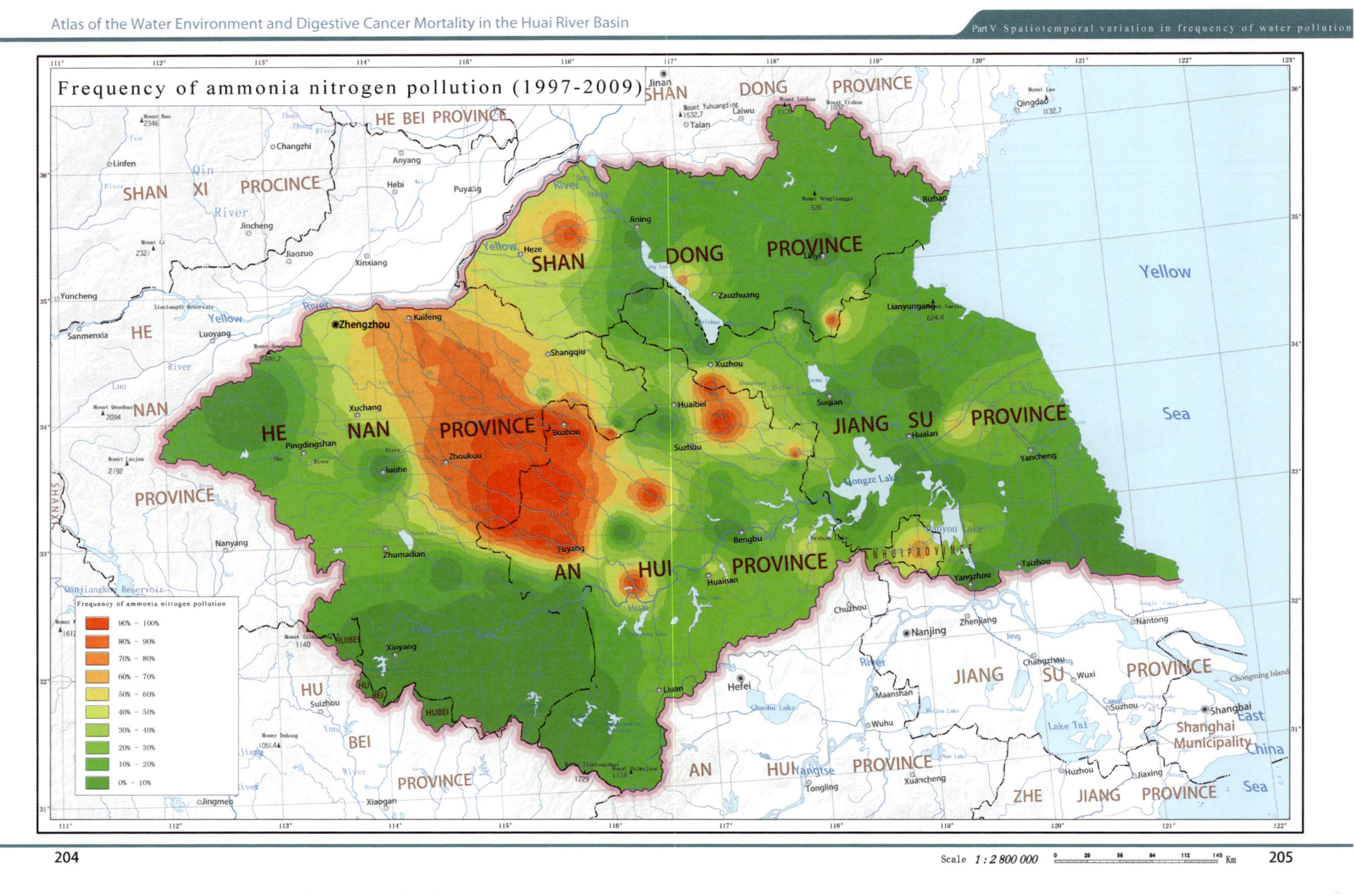

Fig. 6.9 Frequency of ammonia nitrogen pollution (1997–2009)

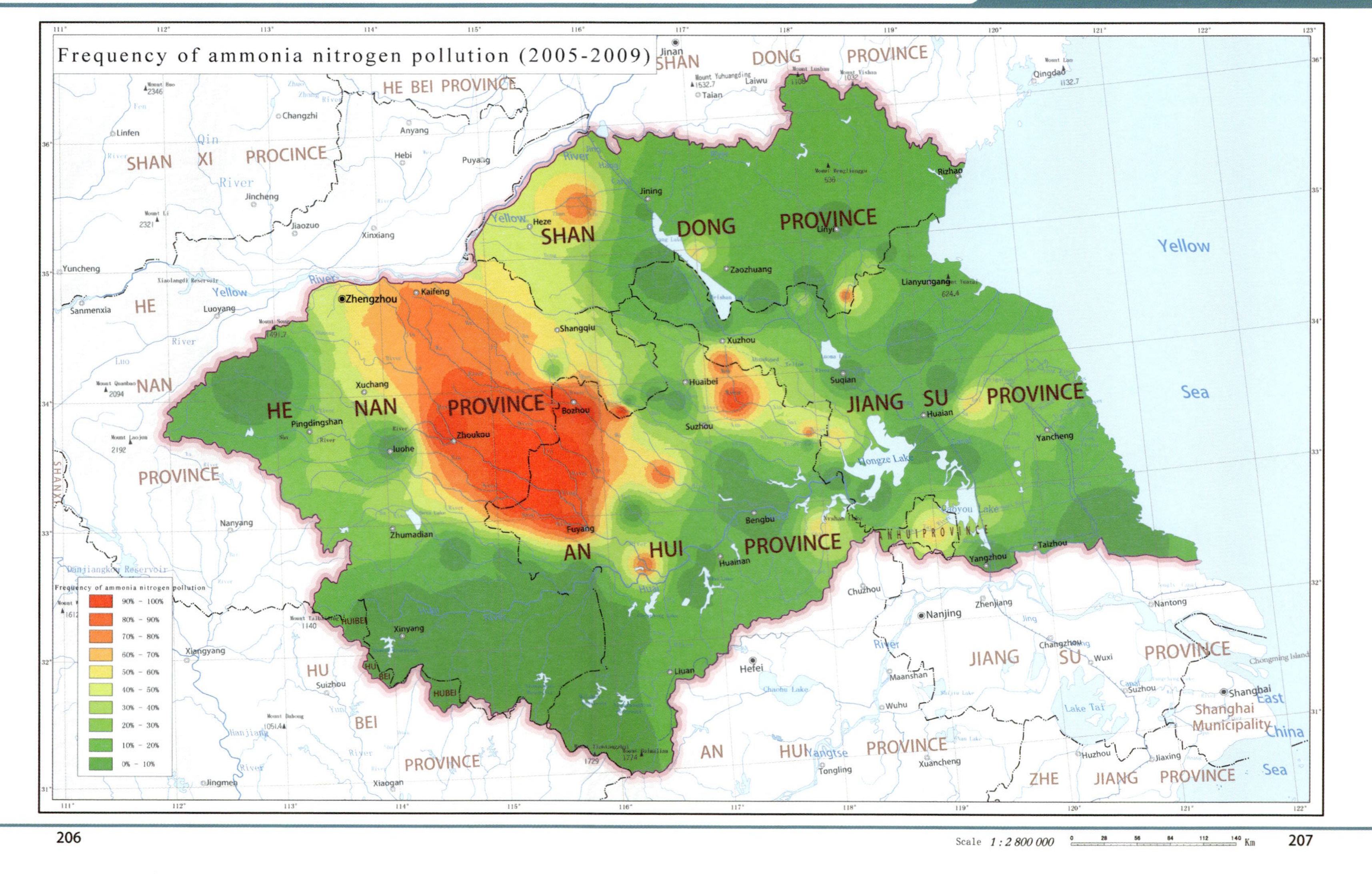

Fig. 6.10 Frequency of ammonia nitrogen pollution (2005–2009)

7 Age-Standardized Mortality Rate of Digestive Cancer

G. Yang, D. Zhuang (eds.), *Atlas of the Huai River Basin Water Environment: Digestive Cancer Mortality*,
DOI 10.1007/978-94-017-8619-5_7, © Springer Science+Business Media Dordrecht 2014

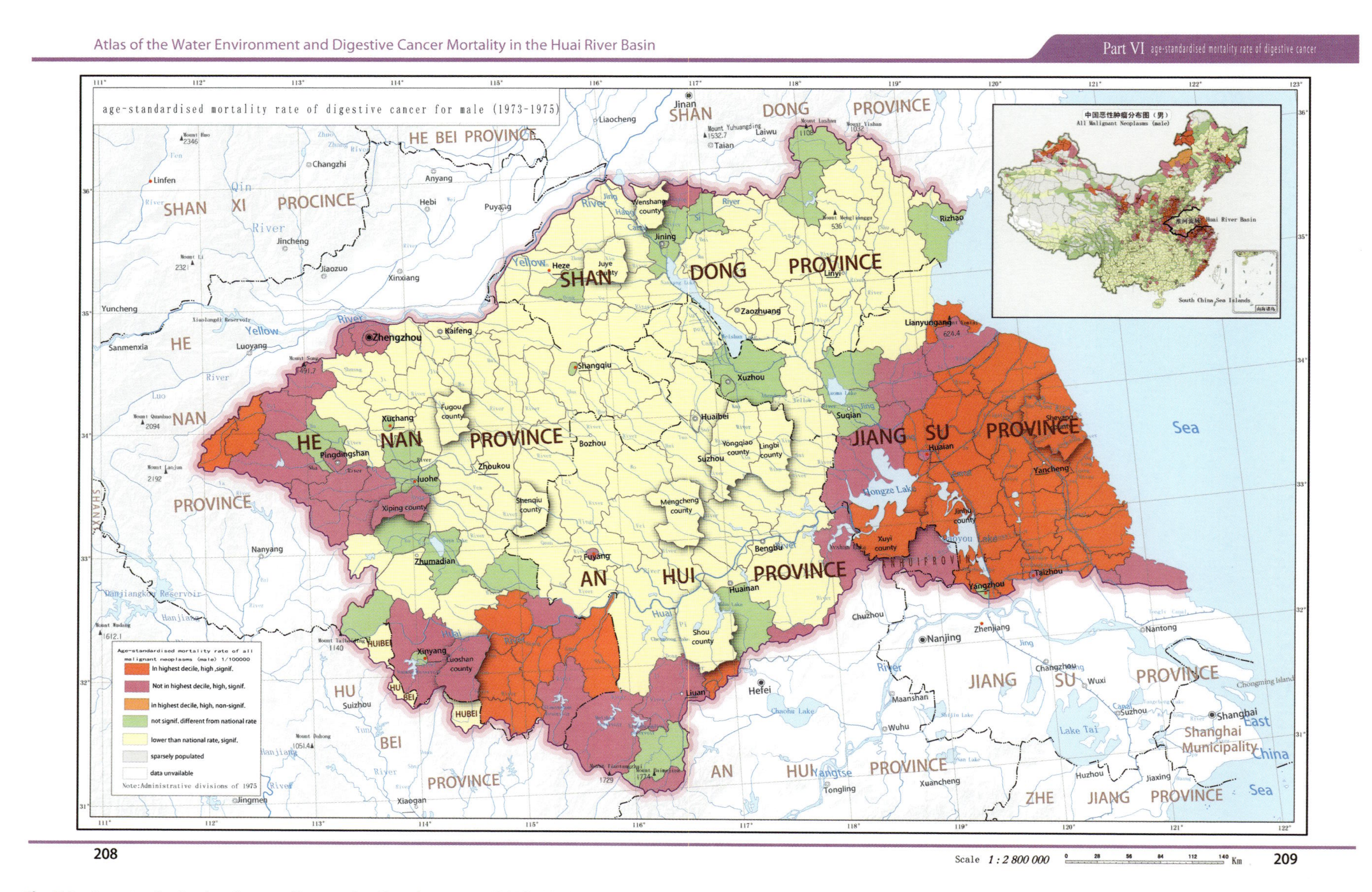

Fig. 7.1 Age-standardised male mortality rate for digestive cancer (1973–1975)

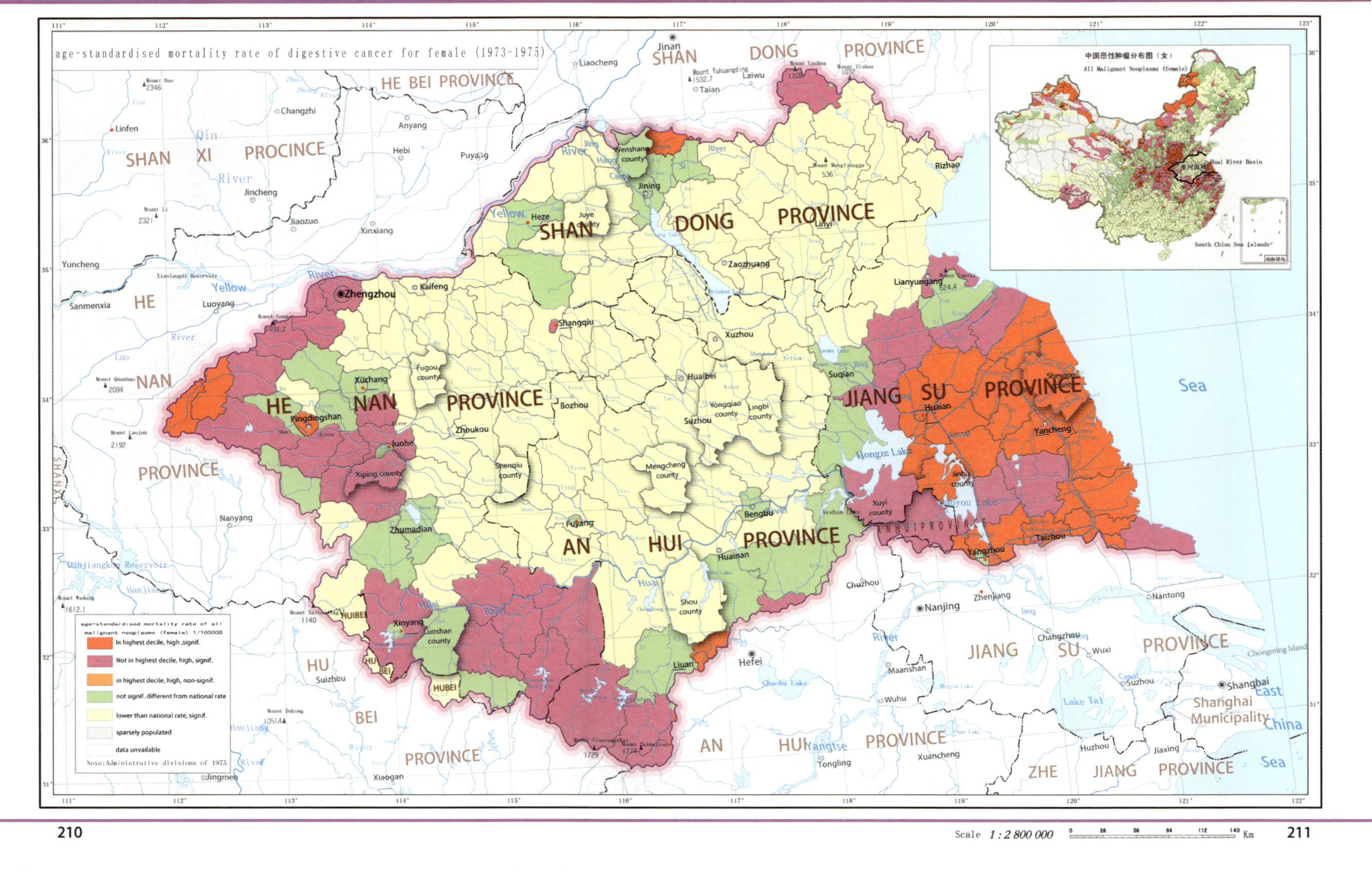

Fig. 7.2 Age-standardised female mortality rate for digestive cancer (1973–1975)

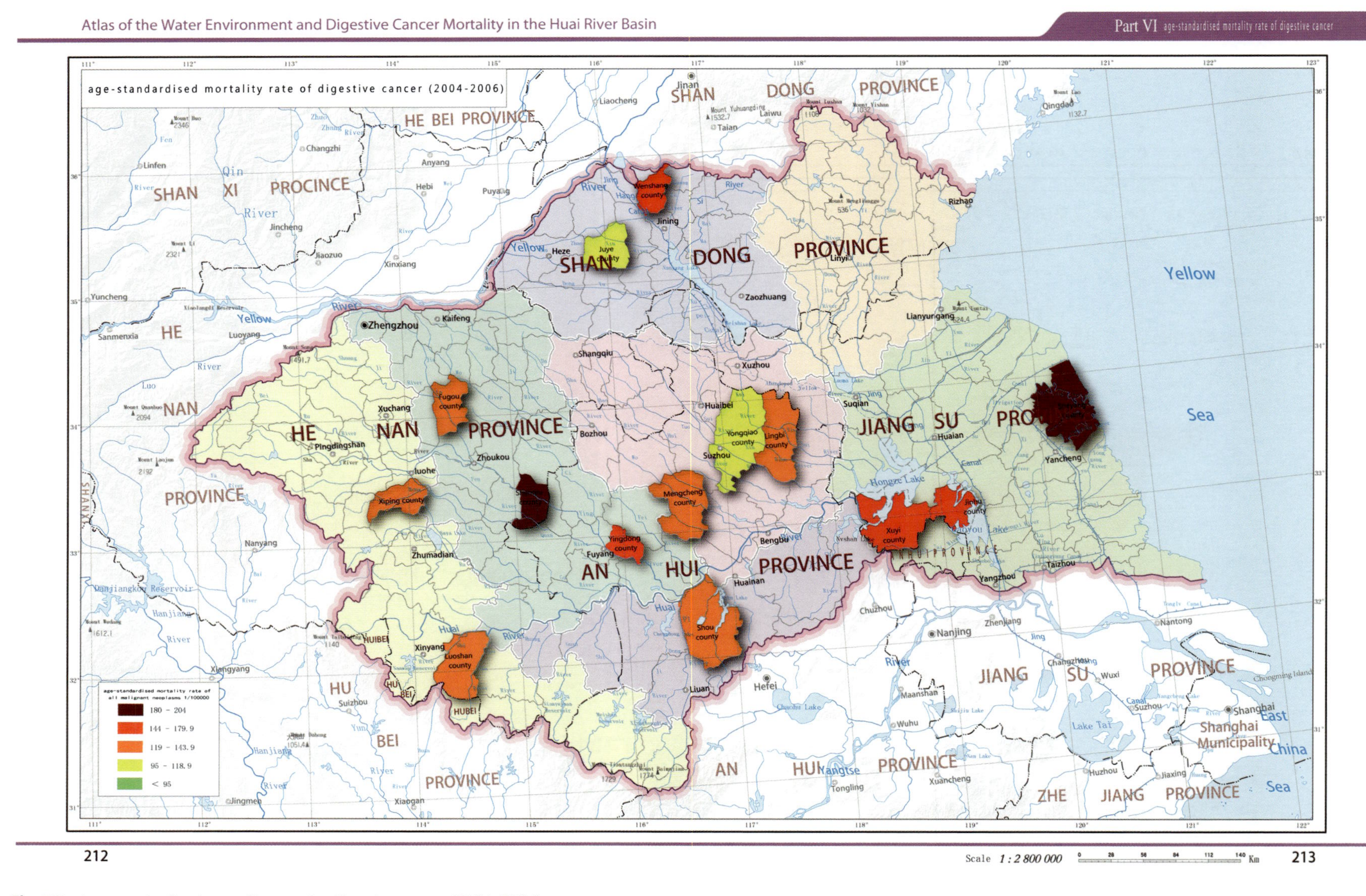

Fig. 7.3 Age-standardised mortality rate for digestive cancer (2004–2006)

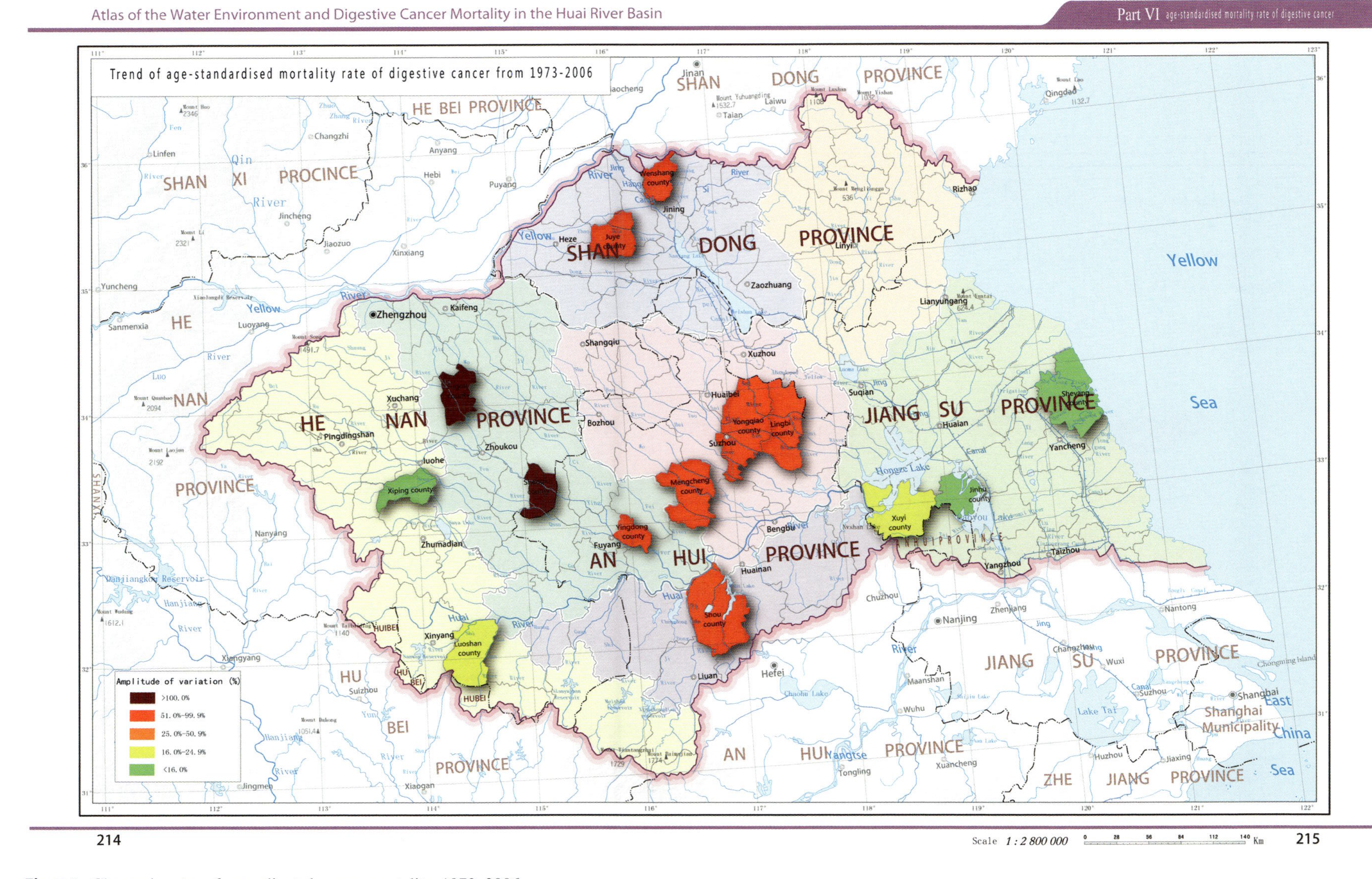

Fig. 7.4 Change in rates of age-adjusted cancer mortality, 1973–2006

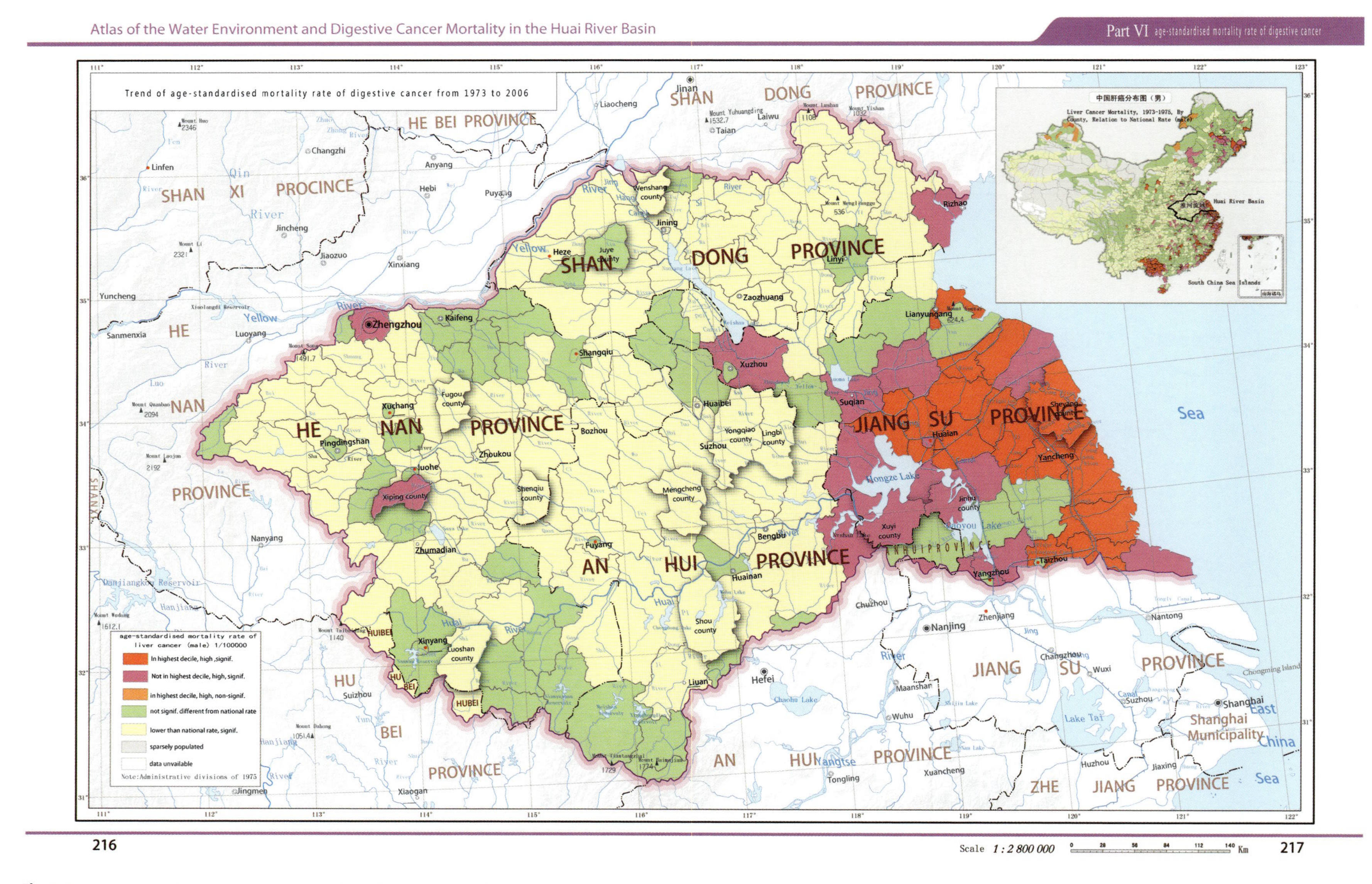

Fig. 7.5 Age-standardised male mortality rate for liver cancer (1973–1975)

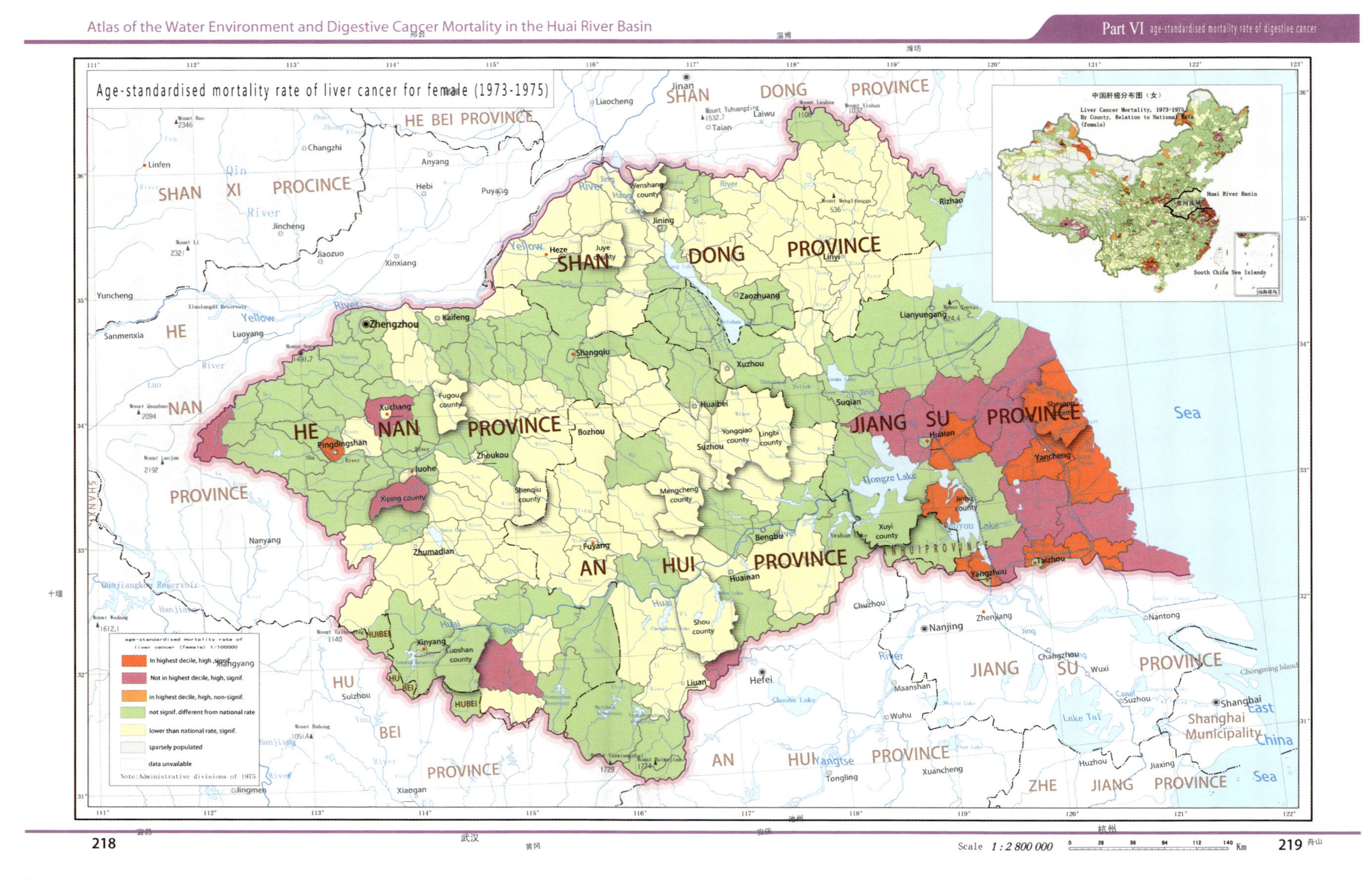

Fig. 7.6 Age-standardised female mortality rate for liver cancer (1973–1975)

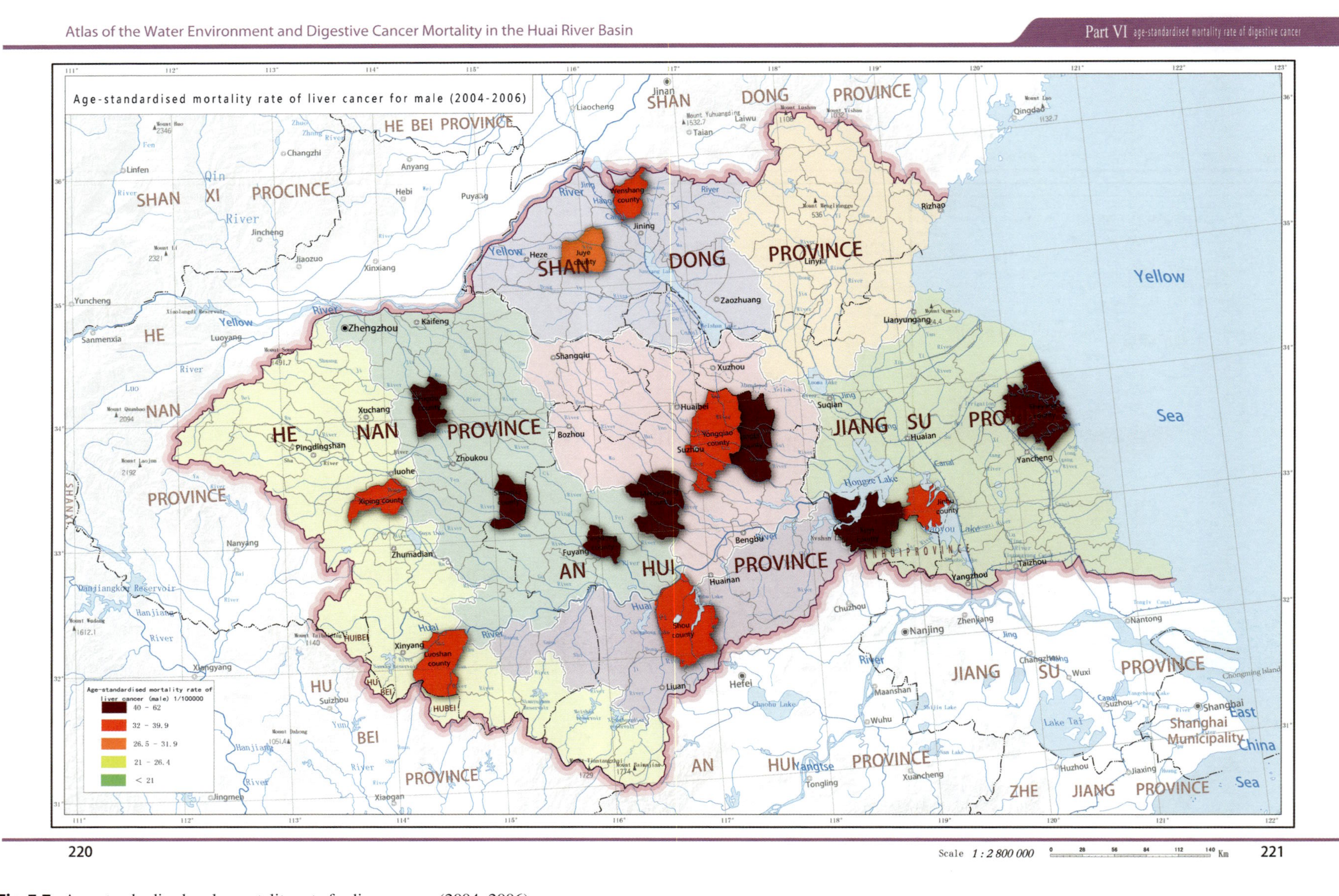

Fig. 7.7 Age-standardised male mortality rate for liver cancer (2004–2006)

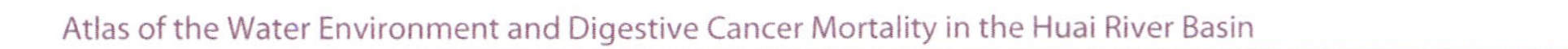

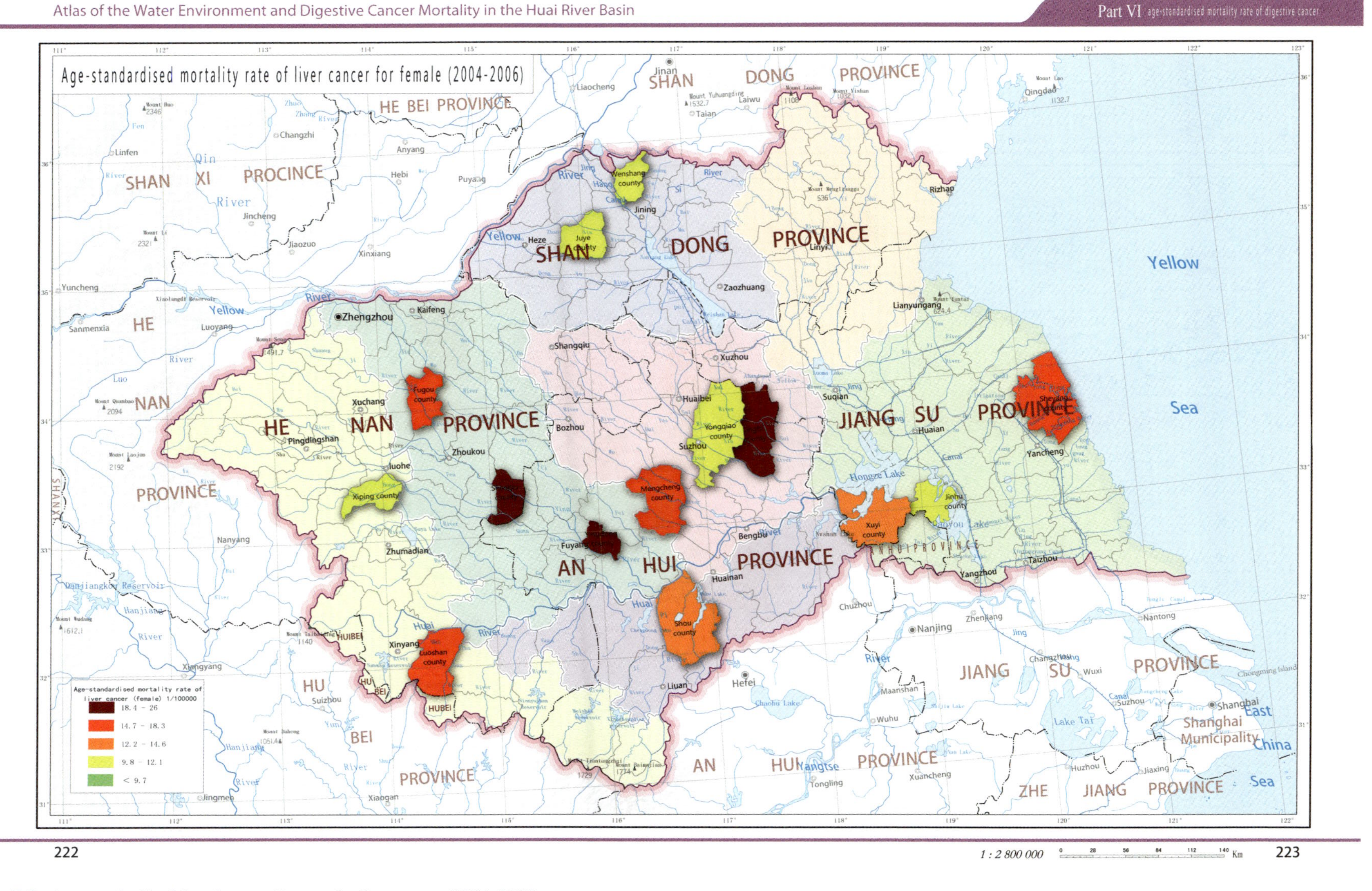

Fig. 7.8 Age-standardised female mortality rate for liver cancer (2004–2006)

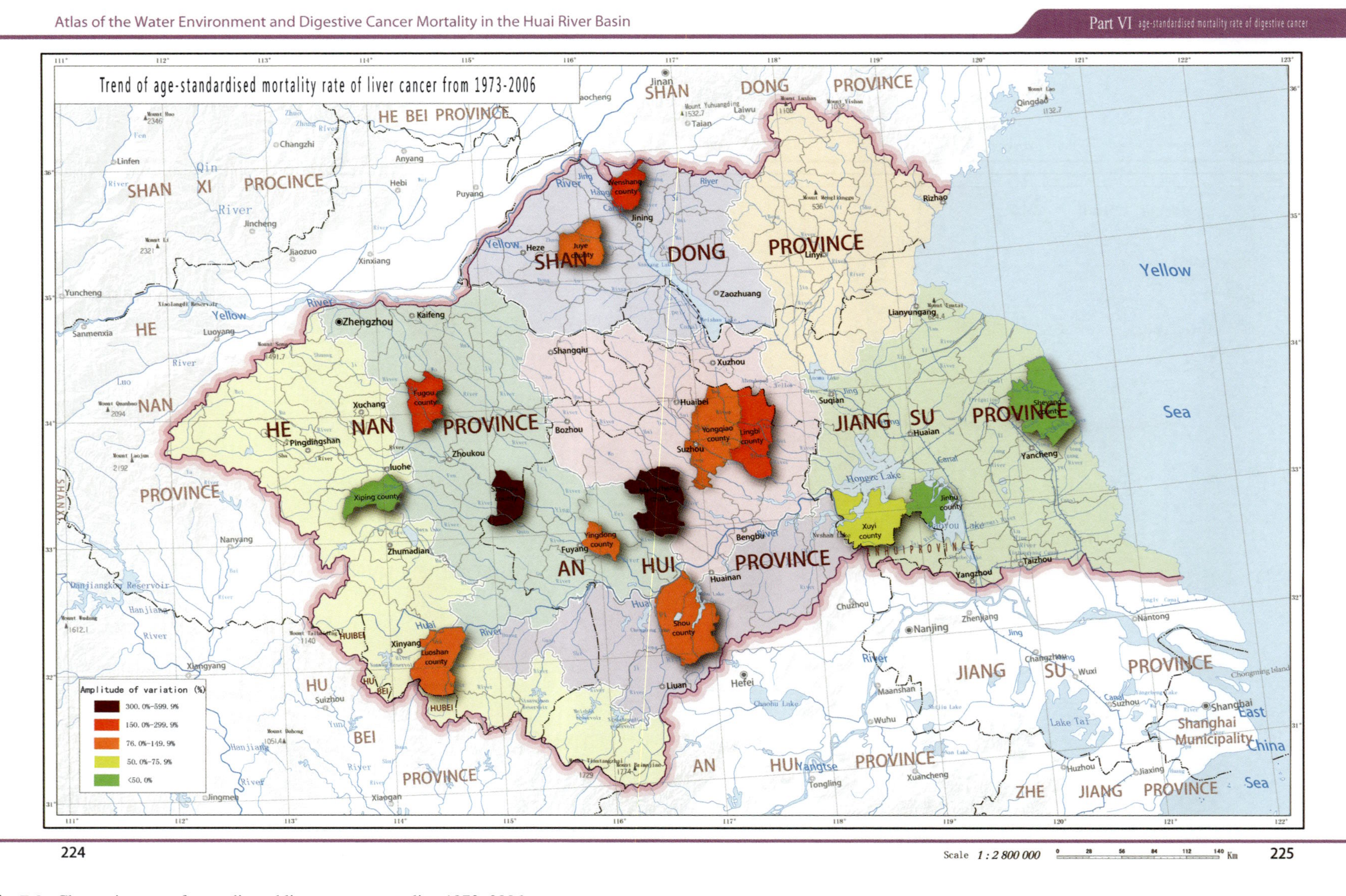

Fig. 7.9 Change in rates of age-adjusted liver cancer mortality, 1973–2006

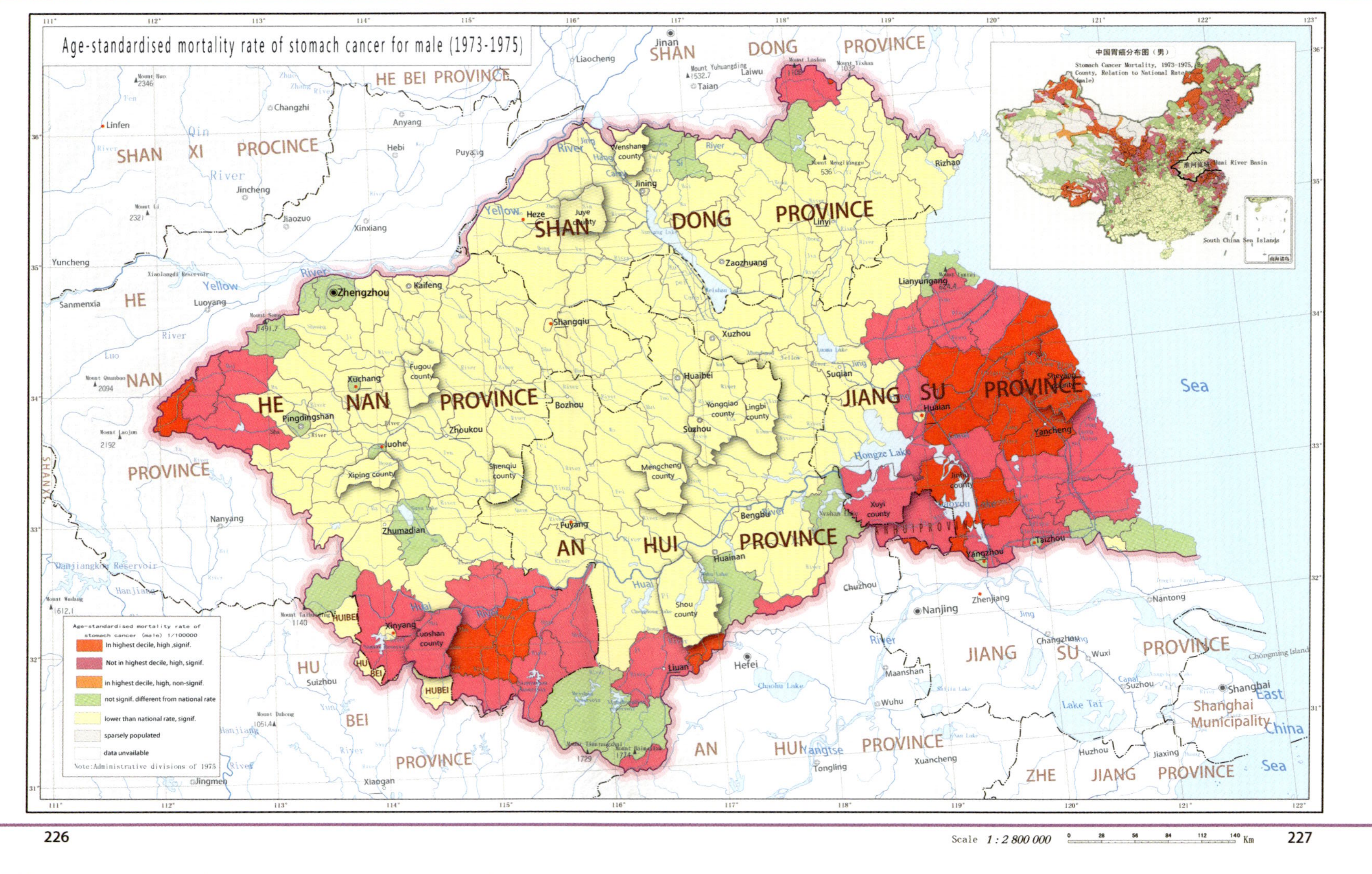

Fig. 7.10 Age-standardised male mortality rate for stomach cancer (1973–1975)

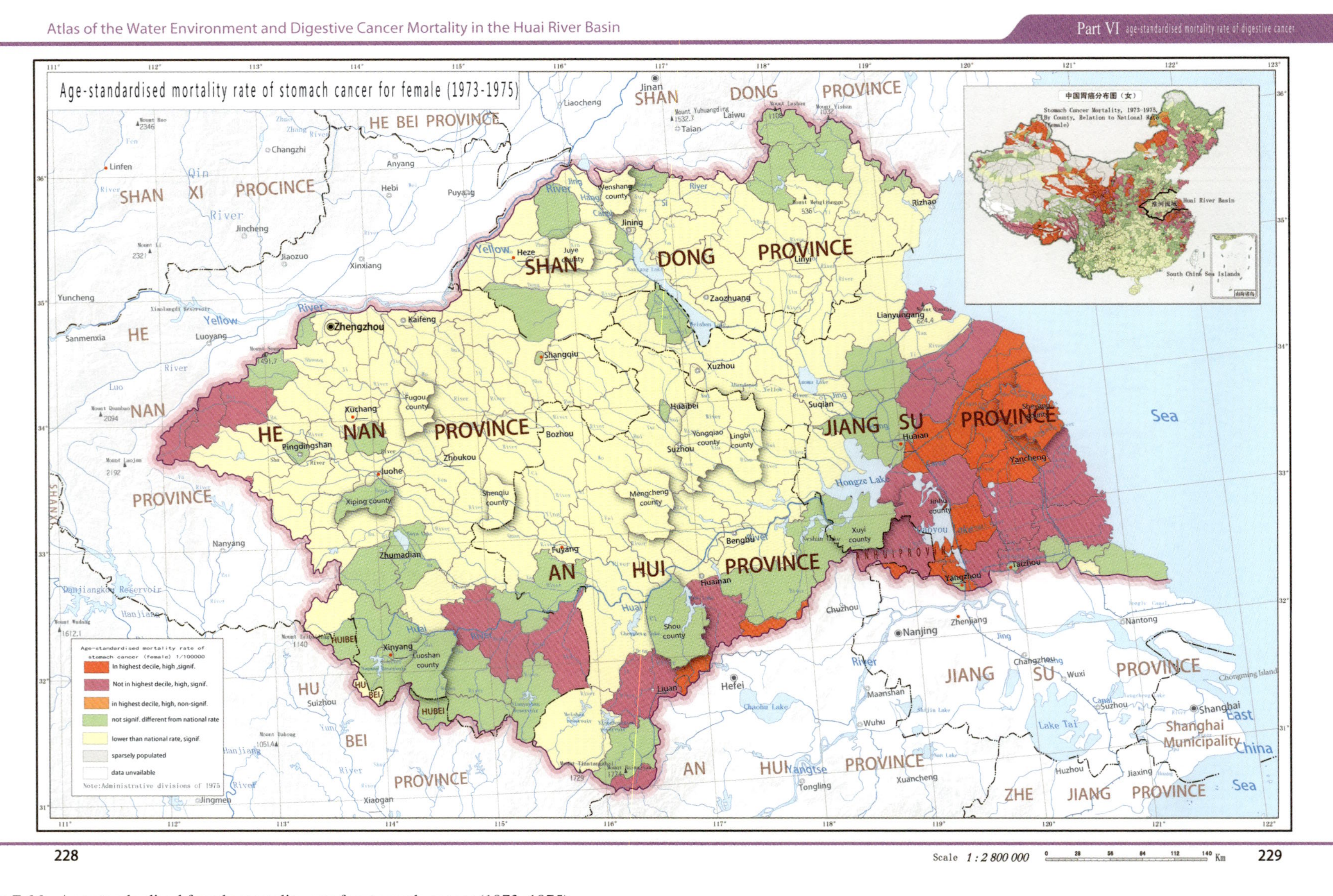

228 229

Fig. 7.11 Age-standardised female mortality rate for stomach cancer (1973–1975)

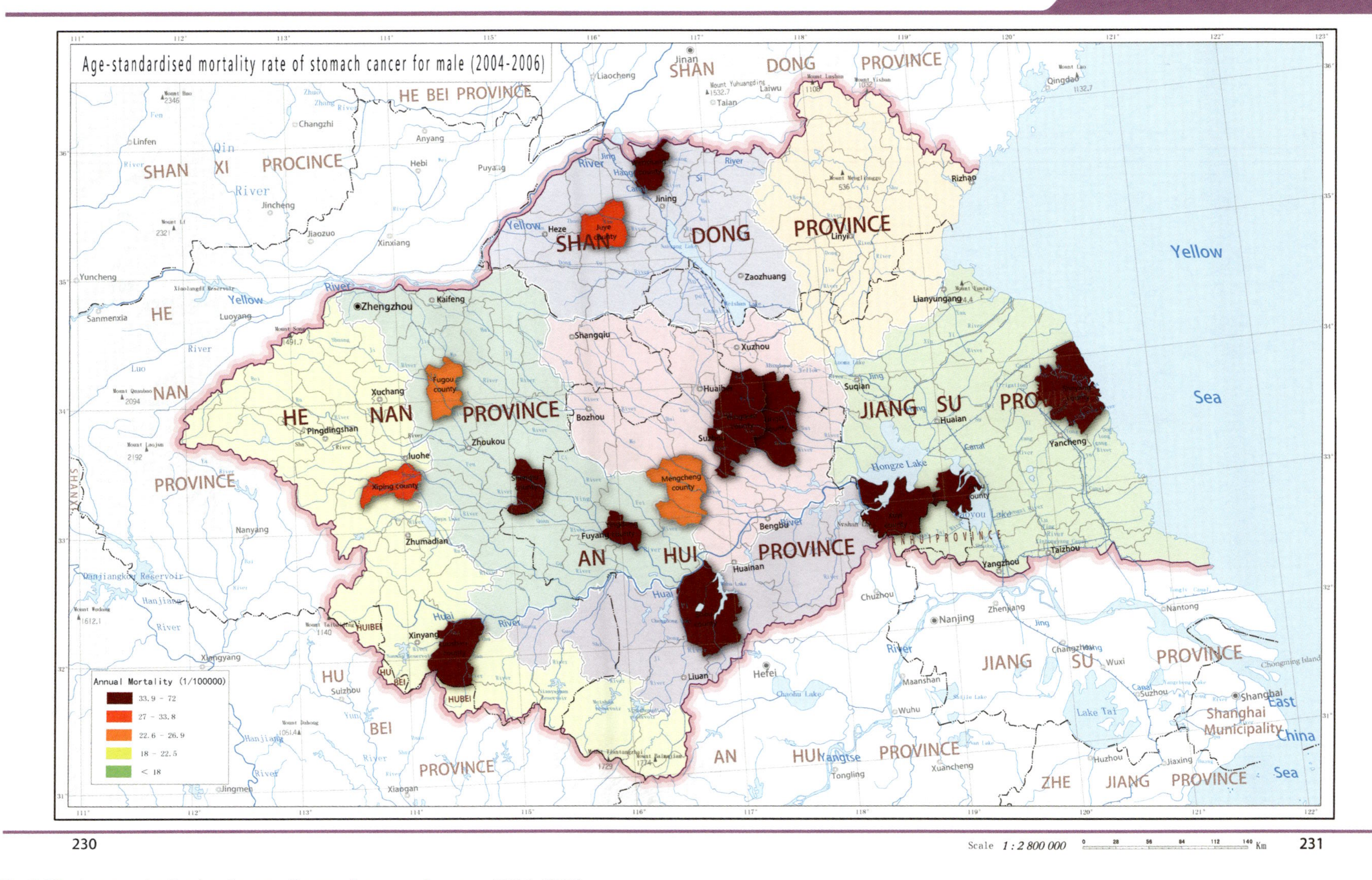

Fig. 7.12 Age-standardised male mortality rate for stomach cancer (2004–2006)

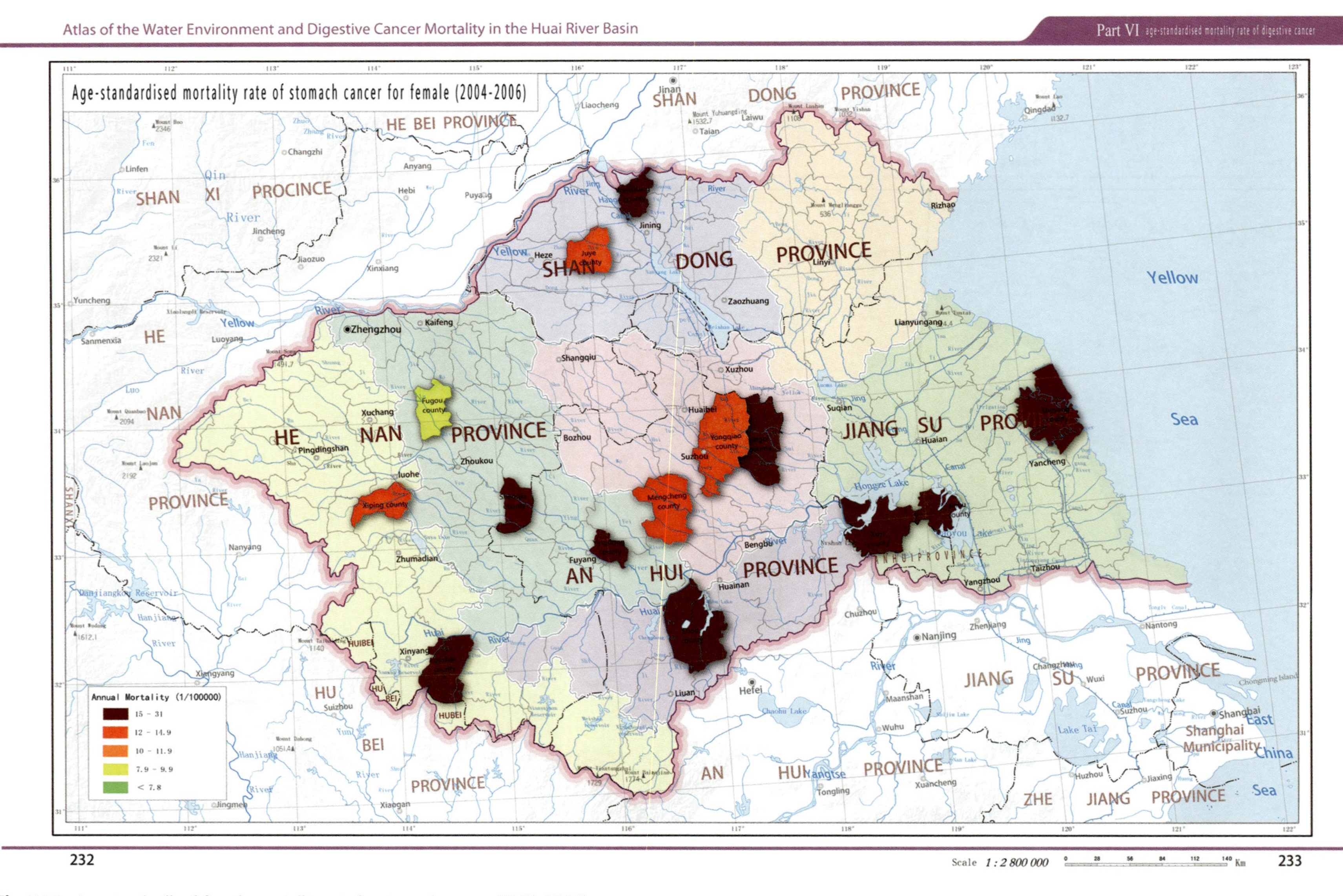

Fig. 7.13 Age-standardised female mortality rate for stomach cancer (2004–2006)

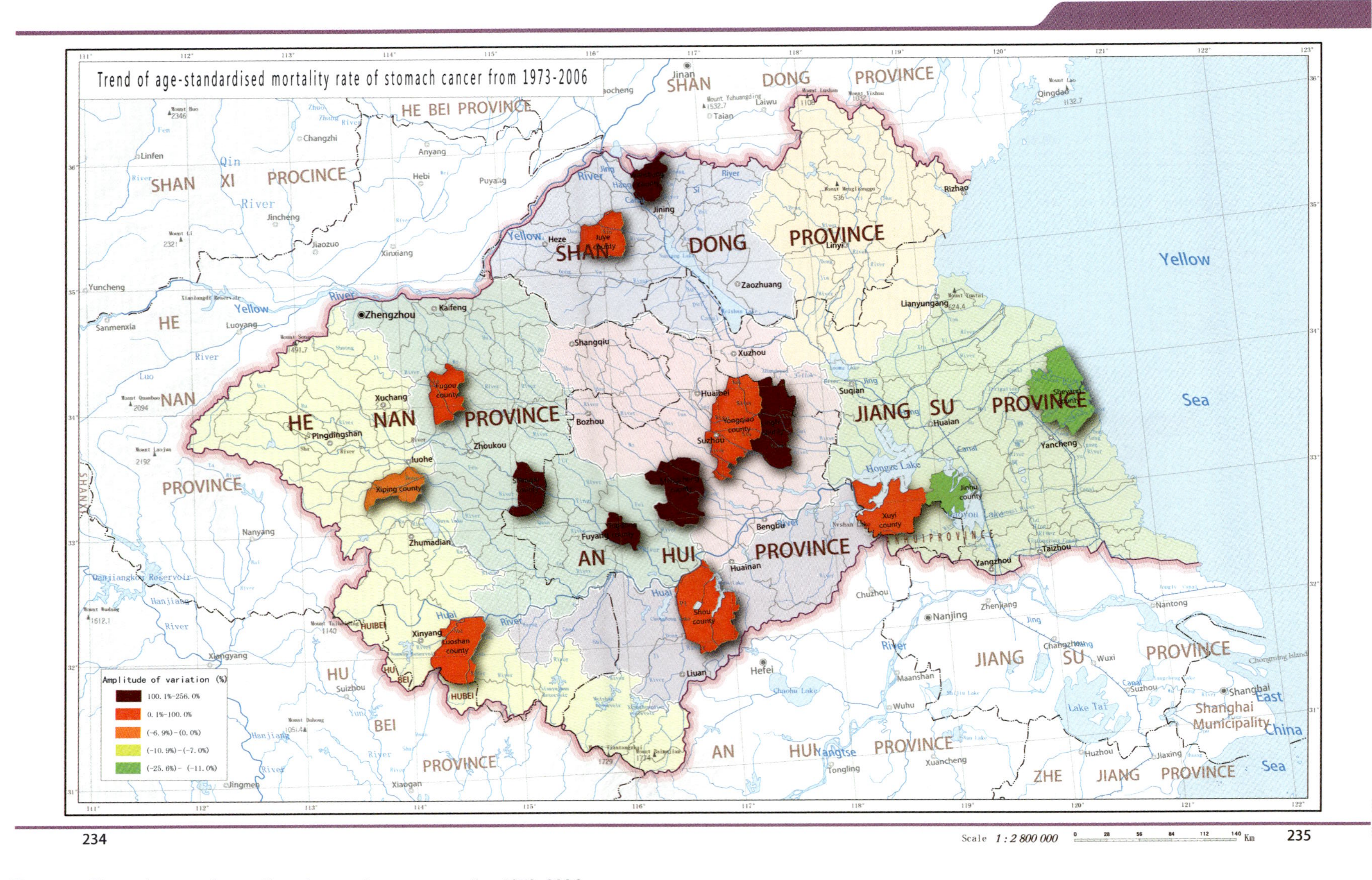

Fig. 7.14 Change in rates of age-adjusted stomach cancer mortality, 1973–2006

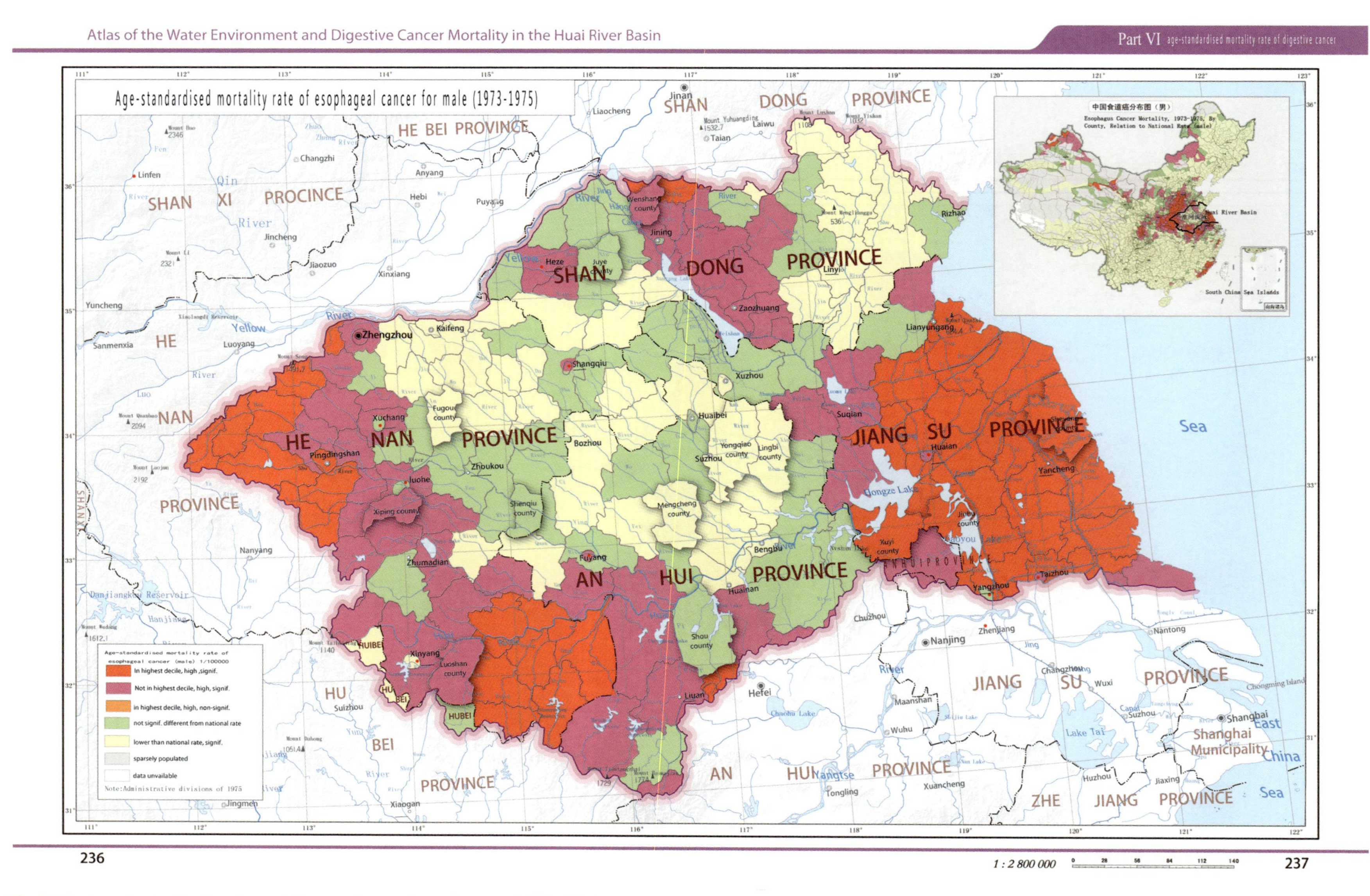

Fig. 7.15 Age-standardised male mortality rate for esophageal cancer (1973–1975)

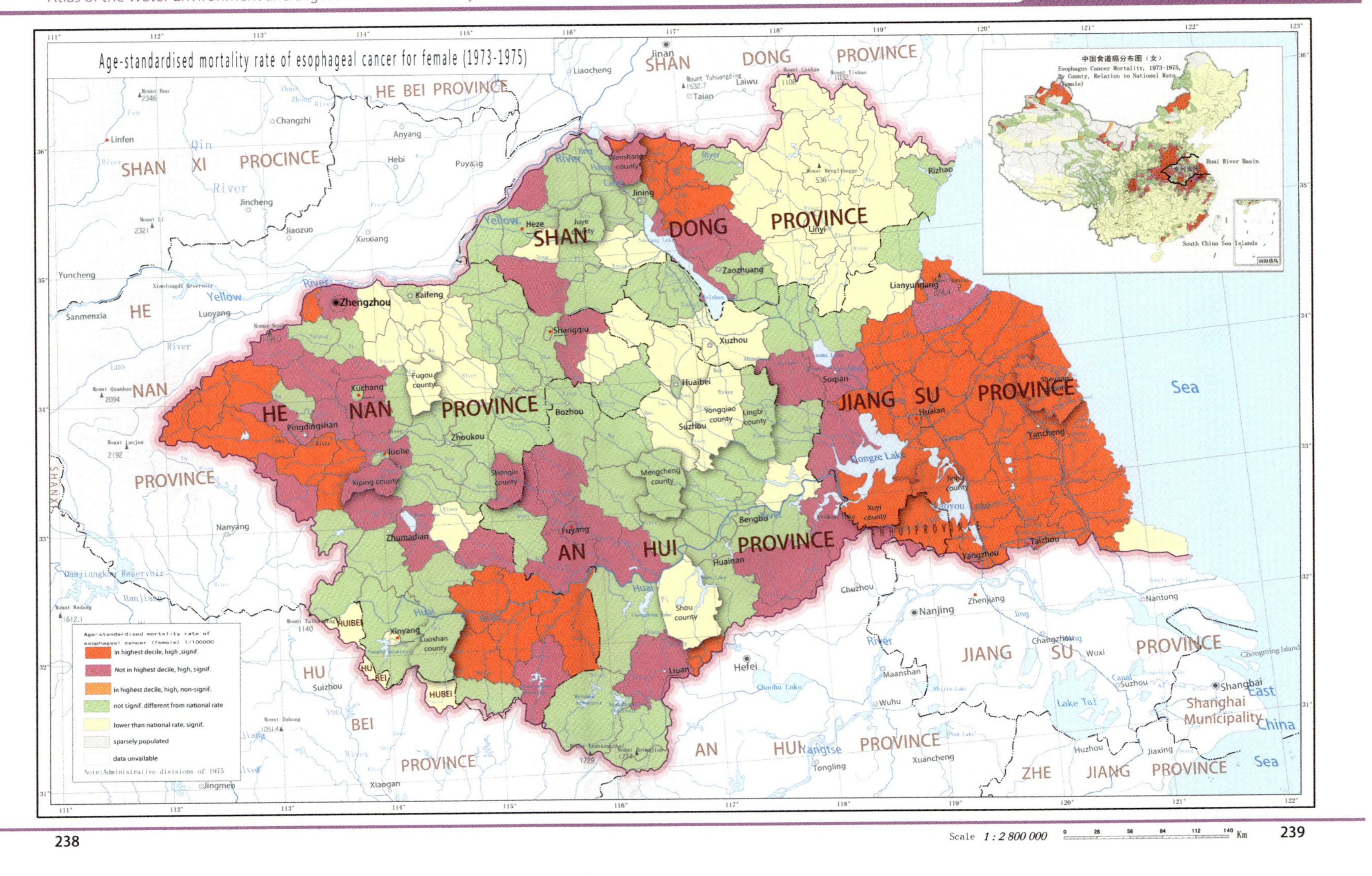

Fig. 7.16 Age-standardised female mortality rate for esophageal cancer (1973–1975)

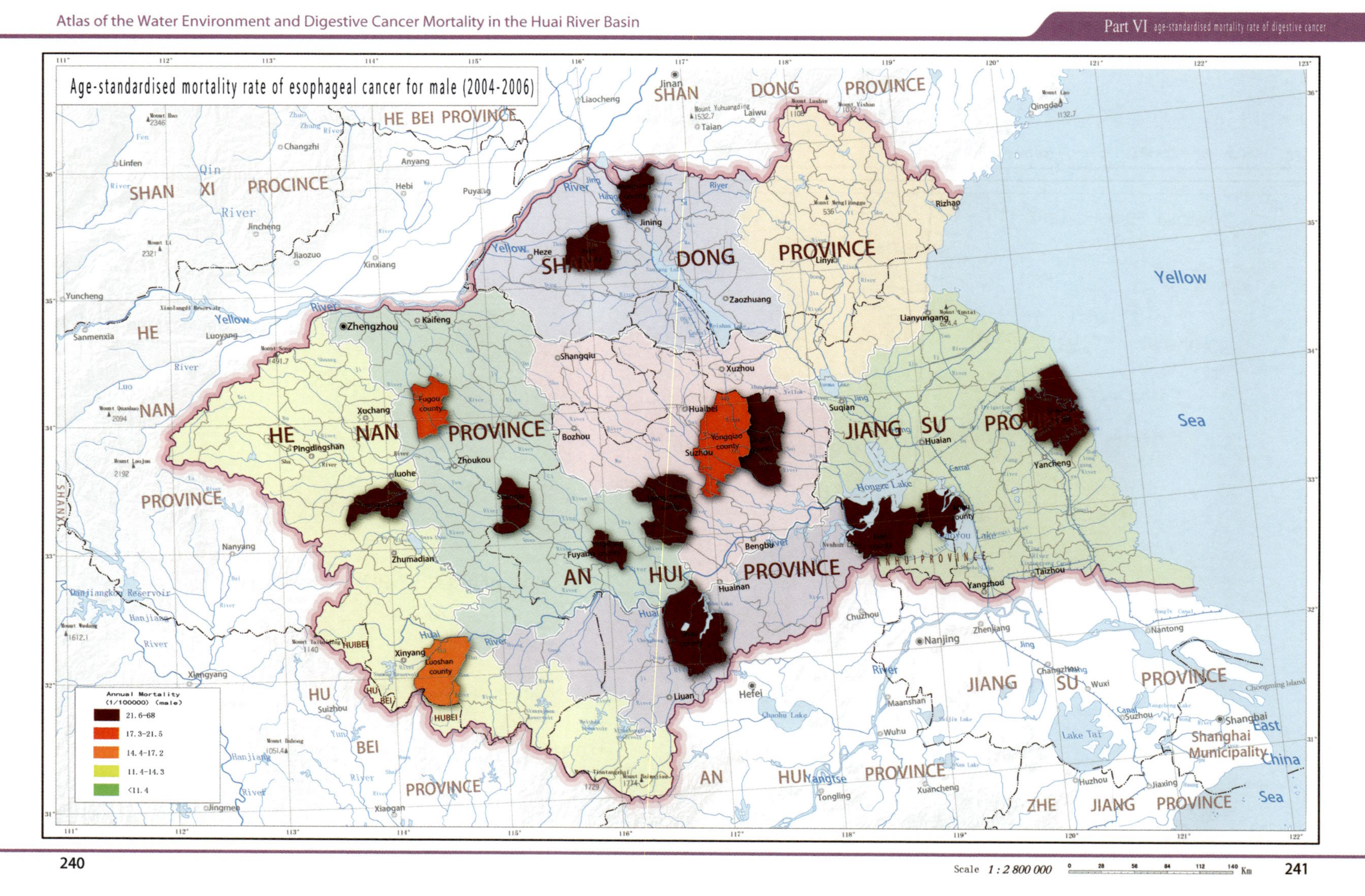

Fig. 7.17 Age-standardised male mortality rate for esophageal cancer (2004–2006)

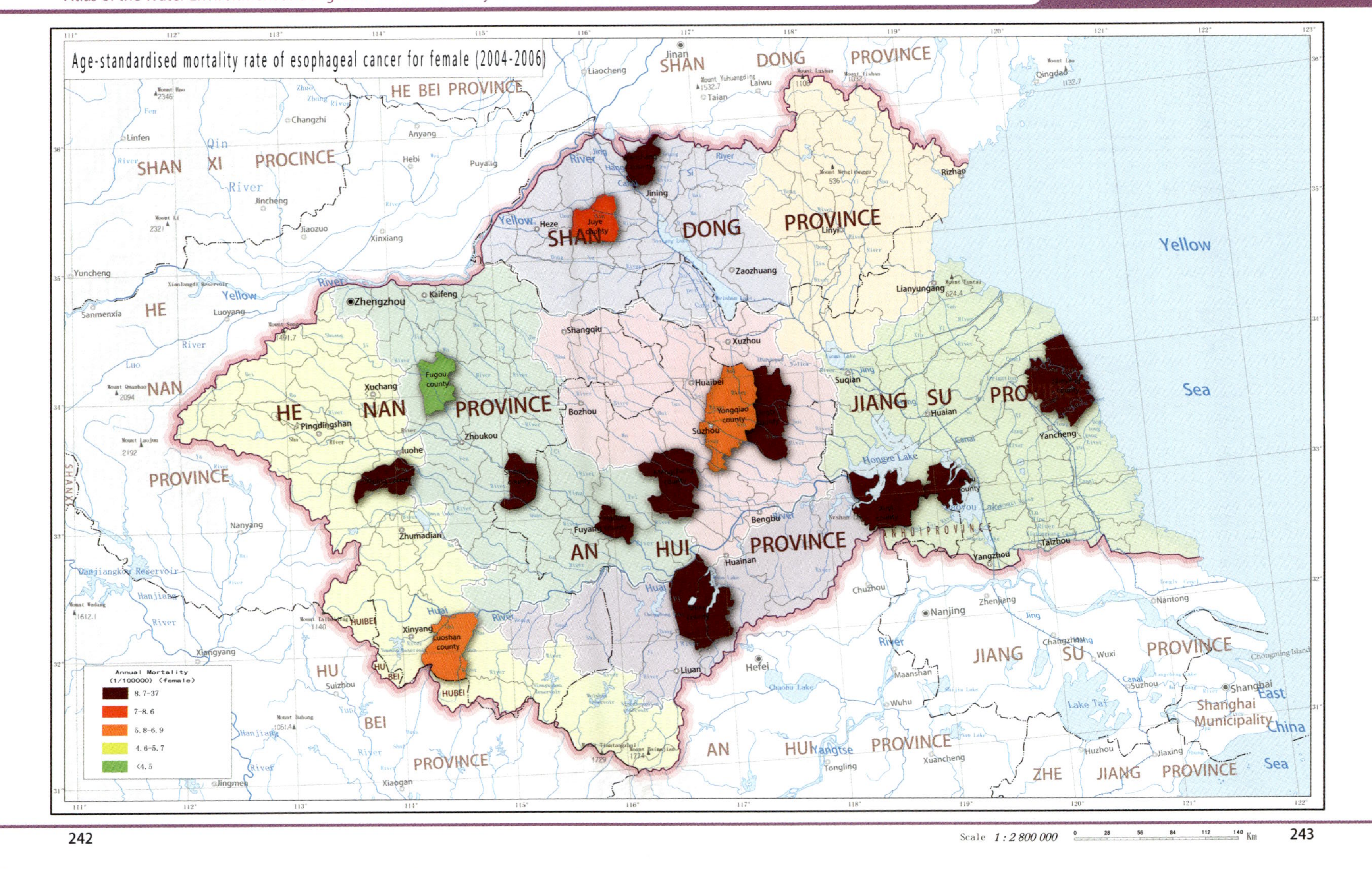

Fig. 7.18 Age-standardised female mortality rate for esophageal cancer (2004–2006)

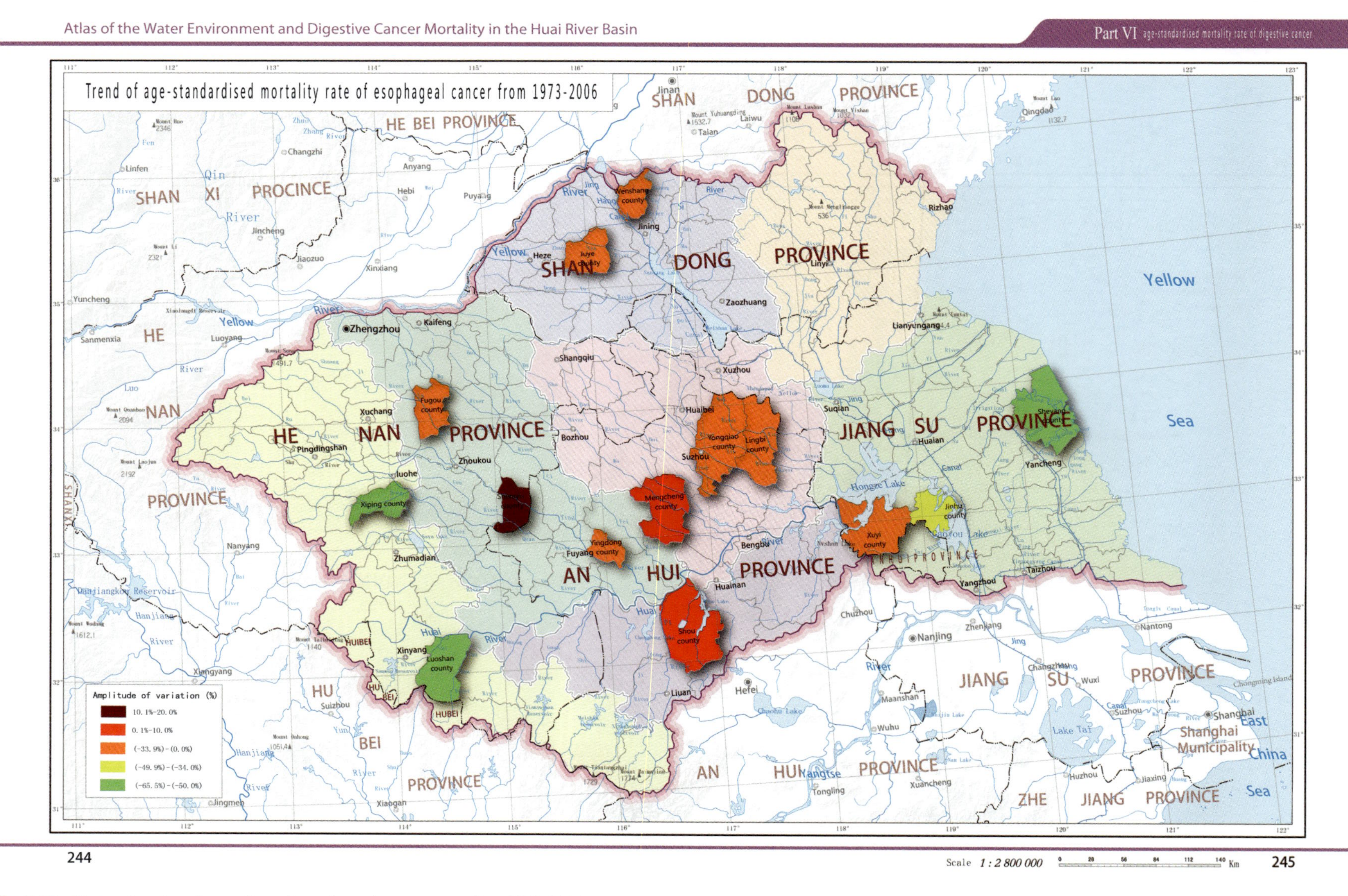

Fig. 7.19 Change in rates of age-adjusted esophageal cancer mortality, 1973–2006

Glossary

Ammonia nitrogen Ammonia nitrogen is considered as an indicator of water quality in GB3838-2002 and specialized for the sum of NH_3 and NH_4^+.

Biochemical oxygen demand (BOD) Biochemical oxygen demand (BOD) is the amount of dissolved oxygen needed to decompose the organic matter in water by microorganisms' biochemical actions.

Carcinoma of stomach Tumors derived from the gastric epithelium are related to exposure to nitrosamines, polycyclic aromatic hydrocarbon compounds, certain dietary factors, helicobacter pylori infection and genetic factors.

Chemical oxygen demand (COD) Chemical oxygen demand (COD) is derived from the volume of oxygen consumed by the oxidizable matters (organic matter, nitrite, ferrite, sulphide, etc.) through oxidation decomposition with chemical oxidants (e.g. potassium permanganate).

Esophagus cancer, carcinoma of the esophagus Esophagus cancer or carcinoma of the esophagus refers to tumors derived from the esophageal squamous epithelium and columnar epithelium.

Frequency of water pollution For monitoring sections with available long-term water quality data, the ratio of the occurrence of water quality Grade V or worse (polluted water) to total observations (frequency) is used to assess water quality. This ratio is defined as the frequency of water pollution (FWP) and calculated via the following formula.

$FWP = Y_p/Y$

Where Y_p is the occurrence of water quality Grade V or worse in the given monitoring period (times); and Y is total observations (e.g. yearly or monthly observations). Here, grade V or worse can refer to water quality after comprehensive evaluation or to the concentration level of a single indicator. Because surveillance data was missing for certain sections for certain years (see Tables 1.4, 1.5, 1.6, 1.7, and 1.8 for details), these years were not taken into account when we calculated these ratios.

Huai River The Huai River originates from Mount Funiu and Mount Tongbai in the west, and flows eastward into the Yellow Sea. The total length of the river is 1,000 kilometers (km) and the total drop is 200 m. The upper reaches of the river, which are above Honghekou, cover a distance of 360 km, with a drop of 178 m and a gradient of 1/2,000. The middle reaches run from Honghekou to Zhongdu with an outlet out into Hongze Lake. They cover a distance of 490 km, with a drop of 16 m and a gradient of 1/33,000. The lower reaches run from Zhongdu to Sanjiangying, covering a distance of 150 km, with a drop of 6 m and a gradient of 1/25,000. In addition to the channel which enters the Yangtze River, there is also the North Jiangsu irrigation canal and the Huaishuxin River, which diverts flood water to the Xinyi River.

Huai River Basin The Huai River Basin is located in eastern China between the Yellow River Basin and the Yangtze River Basin (longitude E 111° 55′ 122° 45′, latitude N 30° 55′ 36° 20′). The Huai River is the natural geographical boundary between China's northern and southern climatic zones. As a transitional zone between the two, it is a warm temperate area with a north Asian, humid to semi-humid monsoon climate and four distinct seasons. To the south it is bordered by the Dabie Mountains, the Jianghuai hills, the Tongyang Canal and the south dike of the Rutai Canal, which separate the Huai River Basin from the Yangtze River Basin. To the north, the Huai River Basin is separated from the Yellow River Basin by the south dike of Yellow River and by Mount Tai. The west, southwest and northeast parts of the river basin are mountainous and hilly, and account for about one-third of the total watershed area. The remaining area is a large plain, which accounts for about two-thirds of the watershed, including lakes and depressions.

Main stream This the main river that collects all the runoff from the catchment in the river system. In this atlas, it refers to the main stream of Huai River.

Malignant neoplasm of bronchus and lung Tumors derived from the bronchial mucosa and glands are related to tobacco use, air pollution and certain occupational factors, including squamous cell carcinoma, adenocarcinoma, adenosquamous carcinoma, undifferentiated carcinoma, bronchial carcinoid and other subtypes.

Malignant neoplasm of the liver Refers to tumors derived from liver cells, and their epidemic is associated with viral hepatitis and cirrhosis, aflatoxin, water pollution and genetic factors.

G. Yang, D. Zhuang (eds.), *Atlas of the Huai River Basin Water Environment: Digestive Cancer Mortality*,
DOI 10.1007/978-94-017-8619-5,

Monitoring results of water quality Monitoring results of water quality are the concentrations or amounts of contaminants in water, which is the primary basis for the assessment of water quality and environmental changes. The scope of monitoring results is widely suitable for uncontaminated or contaminated natural water and all kinds of industrial water. The main indicators of monitoring results are mostly classified into two categories. One is the synthetic indicators reflecting water quality condition, including temperature, chromaticity, turbidity, pH, electro-conductibility, suspended substance, dissolved oxygen, chemical oxygen demand, biochemical oxygen demand, etc. The other one denotes some toxic matters, such as phenol, cyanogens, arsenic, plumbum, chromium, cadmium, mercury, organic pesticide, and so on. Apart from the above indicators, volume and velocity of water flow are also measured in order to assess the water quality situation of rivers.

Monitoring section Monitoring sections, frequently referred in this atlas, are some special points to the special points for the surveillance on water quality of Huai River water system, which are set up and managed by the environmental protection agencies.

Neoplasm, carcinoma Neoplasm or carcinoma mainly refers to malignant tumors characterized by cell differentiation abnormalities, altered proliferation and uncontrolled growth. Carcinoma cells invade surrounding tissues directly or metastasize distantly, involving normal tissues through lymph and blood circulation, and impair their function or lead to their loss, resulting in cachexia.

Nonionic ammonia According to GB3838-1988, nonionic ammonia is one formation of ammonia nitrogen and commonly expressed by NH_3-N.

Per capita gross domestic product As an important economic indicator in development economics, per capita gross domestic product (GDP) is employed to measure the economic development. It is an efficient tool to evaluate the status of national or regional macro-economy. It is equal to the ratio of national or regional gross domestic product in the accounting period (typically 1 year) to the total permanent resident population in this nation or region.

Primary tributary Primary tributaries are rivers that are directly linked to the mainstream.

Secondary tributary Secondary tributaries are rivers that are directly linked to the primary tributaries.

Spatial interpolation Spatial interpolation is a method of converting discrete points to a consecutive surface, which usually aims for the comparison with other spatial patterns.

Volatilizing phenol As a type of monophenols with high toxicity, volatilizing phenol mainly denotes the phenols with a boiling point below 230 °C.

Water quality Water quality is a synthetic assessment for the physical (chromaticity, turbidity, odour, etc.), chemical (concentrations of inorganic and organic matters), and biological (bacteria, microorganism, plankton, benthon) properties and their combinations in water.

Water quality grade In terms of I, II, III, IV, V, and worse than V, water quality grades are derived from the sorted surveillance data of water quality by the criteria of water quality authorized by the environmental protection agencies. In this atlas, GB3838-1988 and GB3838-2002 were respectively applied for the classification.

Index

G. Yang, D. Zhuang (eds.), *Atlas of the Huai River Basin Water Environment: Digestive Cancer Mortality*, DOI 10.1007/978-94-017-8619-5, © Springer Science+Business Media Dordrecht 2014